Science Fair Success

Science Fair Success
Revised and Expanded

Ruth Bombaugh

Enslow Publishers, Inc.

40 Industrial Road	PO Box 38
Box 398	Aldershot
Berkeley Heights, NJ 07922	Hants GU12 6BP
USA	UK

http://www.enslow.com

Library of Congress Cataloging-in-Publication Data

Bombaugh, Ruth.
    Science fair success / Ruth Bombaugh. — Rev. and expanded.
       p. cm. — (Science fair success series)
    Includes bibliographical references and index.
    Summary: A guide for choosing, designing, and completing an investigative
science fair project, with an appendix listing prize winning projects by students.
    ISBN 0-7660-1163-1
    1. Science projects—Handbooks, manuals, etc. 2. Science projects—
Juvenile literature. [1. Science projects.] I. Title. II. Series.
Q182.3.B66   1999
507.8—DC21   98-3297
                                       CIP
                                       AC

Printed in the United States of America

10 9 8 7 6 5 4 3 2

To Our Readers:
All Internet addresses in this book were active and appropriate when we went to press.
Any comments or suggestions can be sent by e-mail to Comments@enslow.com or to
the address on the back cover.

Illustration Credits: Charles Bombaugh, p. 62; © Corel Corporation, pp. 7, 16, 28,
38, 49, 57, 66; C. Ruth Neudahl, pp. 9, 13, 19, 23, 24, 25, 30, 35, 39, 59, 60, 64,
67, 69, 72; Roy Luck, pp. 21, 34.

Cover Illustration: The Stock Market

Contents

Foreword

Congratulations! Your decision to do a science fair project shows that you are a special student. You are willing to work hard and take risks. Hang on to that confidence in yourself! You *can* win honor for yourself, your family, your school, and your community. Even more important, the knowledge you gain as you strive to win this honor will serve you well throughout your academic career and in your adult life.

I have helped thousands of middle school students with their science fair projects. These students were all "first-timers" just like you. Sometimes a first-timer can be at a serious disadvantage compared with students who have already had experience competing at science fairs. Even so, many of these greenhorns won top honors for their work. Some of these students came from wealthy homes, some from welfare homes. Some were labeled "gifted" by the school, and some were labeled "learning disabled." What determined which student did the best job and won the most honors was *not* whether it was his or her first science fair or third, *not* what kind of home he or she came from, and *not* how "smart" he or she was. Rather, the top honors were awarded to the students who had done the best job, and these students were the ones *willing to work hard and follow through.* Believe me, I know this is true. I have seen it happen over and over again: Work hard and you will succeed.

This book is based on the guidance I have given my students. That means thousands of students before you have been my "test cases"; they have helped me find out what step-by-step guidelines work best for putting together an excellent science fair project. You are the beneficiary of these experiences. You are also benefiting from the many more thousands who read the first edition of this book and who provided the impetus necessary for its revision and expansion. I owe them a debt of gratitude, too.

Ready to get started? Let's begin with "Step 1: Getting Organized for the Tasks Ahead."

Step 1

Getting Organized for the Tasks Ahead

When you decide to do a science fair project, you are making a commitment of time, energy, and perseverance that is unlike, probably, any commitment you have ever made before. It is not an easy commitment to make; many more students start science fair projects than finish them. Somewhere in the middle of the project, after the freshness has worn off, it is easy to get bogged down. A first-time participant, especially, can be overwhelmed by the multitude of details and tasks that need to be accomplished. Here are two ways you can prepare yourself mentally to meet the challenges ahead: (1) Realize that there will be a certain amount of drudgery, and do not let it discourage you. (2) Leave yourself ten times the amount of time you think the project will take. You will need the extra time because unforeseen problems will inevitably pop up. You will be able to resolve these problems, but only if you leave yourself enough time to work through them.

You will never find out how much fun a fair can be unless you actually get there. At the fair, you will be able to share your project with fellow students as well as with the judges, and you will all learn from each other. The fair is for you, not just for the judges. The pride and excitement you will feel will soon make you forget how tough it was to get the work accomplished. You have to finish your project first, though, so do not let yourself get discouraged, and leave yourself lots of time!

Now that you know you will have to work hard, you need to find out what kind of work you will be doing to put together a project. You will also need an approximate timetable. The rest of this chapter is an overview of the steps that go into completing a science fair project. It will give you a picture of the work and time required before you get started on your project; it is a good way of getting mentally organized for the tasks ahead.

Getting Started

For many people, the term *science fair project* calls up a mental picture of a three-sided cardboard backdrop decked out with bright colors. Perhaps that is why most students who do a project for the first time think that they ought to start by planning a visual display. In fact, the visual display should be one of the *last* things you do on your project. You will need to have all your data collected, pictures taken, and reports written before you will be ready to think about the best way to display them.

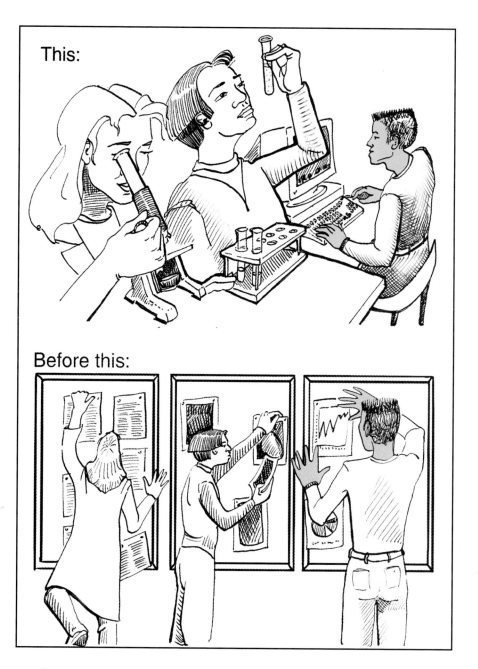

This:

Before this:

For many people, the term *science fair project* calls up a mental picture of a three-sided backdrop. The backdrop, however, is one of the last parts of the project on which you will work. Performing an experiment and writing a research report come first.

So how should you start? First of all, make sure you have chosen the best topic possible. By "best" I mean best for you. Remember that you will be spending a great deal of time and effort on this project; be sure your topic is one you can get excited about. Suggestions in "Step 2: Choosing Your Topic" should help you decide where to look. You will also find examples throughout the book of prizewinning science fair projects that have been submitted by first-time entrants in past years. These projects should help inspire you. Finally, you can look in the back of this book in "Appendix A: Prizewinning Science Fair Projects." There you will find well over a hundred brief suggestions for science fair experiments. You could adapt any one of these ideas to use in a project of your own.

Designing Your Experiment

Once you have selected your topic, you will need to design an experiment that is related to it. The experiment is the single most important part of your science fair project. Performing an experiment in which you collect data and come to a conclusion makes your project scientific rather than, say, critical, like a language arts or social studies project.

Your experiment will be designed to answer a question that you come up with. For example, you might want to find out whether a certain kind of fish would grow faster if it were put in a bigger aquarium, or you might want to find out whether there are some students in your class who are at risk for having heart disease later on in life. There are multitudes of questions that need to be asked by someone like you! "Step 3: Designing Your Experiment" should help you think of a good research question.

The next step in designing your experiment may seem backward or even wrong. You must make your best educated guess of what the answer to your question will be before you start to actually do your experiment. That educated guess has a name; it is called the hypothesis. It will be based on reading that you have done and any background knowledge that you happen to have about the subject. It is not important whether the hypothesis is wrong or right, because that is for the experiment to answer. It is important, however, that you make a hypothesis. Your hypothesis will help guide the development of your experiment through the steps that follow.

You will also need a list of the materials required for the experiment and a step-by-step description of how you will go about doing it. You need to put a lot of thought into this description, which is called the procedure, so that the results of your experiment really answer the question you asked.

Once you have designed your project, you will have all the parts necessary for a proposal: topic, question, hypothesis, materials, and procedure. Then, you will actually set up your experiment and gather the data from it. By carefully looking at your data, you can answer the question, "Is the hypothesis correct?" The answer to this question is the conclusion of your experiment. Then you will have everything you need to make a summary, including the results (in graph or chart form) and conclusion. Both the summary and data keeping are explained in "Step 4: Keeping Track of Your Experiment."

Writing Your Research Report

Some experiments take weeks, months, or even a year to finish because they require enough data to make the results believable. While you are collecting your data, you should also be working on a research report. The research you do at the library in order to write this report will help you understand how your question fits into the broader field of your topic. It will also help when the judges ask you questions. Your report is an integral part of the project, and in most cases the judges will have read the research report even before the judging starts. "Step 5: Writing Your Research Report" will give you ideas on what information to include in your report and on how to write your bibliography.

Putting Together a Visual Display and a Speech

After your data is collected and your research report is written, you will be ready to work on an oral presentation and a visual backdrop. These are the showmanship parts of your project. In essence, you need to "sell" your project to the judges so that they can appreciate the hard work that you put into it and can see the results that you obtained. Information on visual backdrops can be found in "Step 6: Making an Attractive Visual Display." Pointers on how to organize your oral presentation can be found in "Step 7: Talking to the Judges."

Using a Computer to Enhance Your Project

Do not hesitate to log on to a computer. Use powerful search engines to look up background material on the Internet, word-process your research report, and organize your data using spreadsheets or data banks. Need help using any of the software? There should be people willing to give you a hand at the local library or your school. It is worth the extra time

Log on to a computer to enhance your project! Use powerful search engines on the Internet to access background information, organize your raw data onto spreadsheets, and save hours of editing by using a word-processing program. Need help? Go to your teachers, librarians, or other computer-literate friends.

to learn these programs. Students using computers have the advantage of a powerful tool that turns out work more efficiently and accurately than paper and pencil. For the best possible results, make sure that you are one of these students!

Keeping a Logbook

You need to keep a logbook, or journal. Any standard-sized notebook will do, but you should be sure that you do not use it for anything else. In the logbook, record the date, the type of work you did on your project, and how much time you spent doing the work. For example, after today's date, you could record how much time you spent reading this book.

You also will keep your "raw data" in your logbook. The raw data is the first recording of measurements or observations that you make on your experiment. Never throw out your raw data even if it looks sloppily written or is full of jelly spills. It is very important that you keep it *just the way it is.* Some judges will ask to see the raw data in your logbook, so be sure to have it available when you go to the fair.

Planning with a Schedule

Now that you have seen a project overview, maybe you can better appreciate why you should try to leave yourself as much time as possible. You *can* resolve unforeseen problems, but only if you have left yourself enough time to deal with them.

Although no single time schedule would fit all science fair projects, the rough ten-week schedule on the next page has worked for many first-timers.

Good luck on your science fair project. The hundreds of thousands of students who have finished science fair projects before you are testimony to the wonderful work that young people can do in science. So go ahead: Get your logbook started, and proceed to "Step 2: Choosing Your Topic."

Week 1 Choose your topic.

Week 2 Write a proposal for your experiment.

Week 3 Gather the necessary materials and set up your experiment.

Week 4 Start your experiment and be sure to record the raw data in your logbook.

Week 5 Continue recording data on the experiment. Start working on your research report by locating pertinent information at the library and writing a bibliography.

Week 6 Continue recording data on the experiment. Take notes for the background information part of your research report.

Week 7 Continue recording data on the experiment. Write the background information part of your report.

Week 8 Continue recording data on the experiment. Write the further explanation part of your report.

Week 9 Organize data into charts or graphs. Add the results and conclusion parts to the proposal to complete a summary of the experiment. Write an abstract for the project. Complete the report by adding a table of contents and a title page.

Week 10 Make the visual backdrop and practice for the oral presentation.

Step 2

Choosing Your Topic

There is only one limiting factor for choosing your topic: *You must be able to come up with an experiment related to that topic.* It would be difficult, for example, for a student living in Ohio to do a project on birch trees or baleen whales. Birch trees are rare in Ohio, and there are no whales in Ohio except at Sea World. The chances of being able to design an experiment and collect data on these organisms, then, would be slim. Whales and birch trees aside, however, there are a multitude of possible topics surrounding you; just reach out and grab one!

How can you figure out which one to grab? Well, this chapter will give you three different surefire methods students have used. Try any of them. Better yet, try a combination of them. Choosing your science fair topic should be a very personal choice; you need a science fair project that is custom fitted to you. Making a good choice is the first big challenge you have to meet in doing your project. You do not want a topic that is so difficult that you set yourself up for failure before you even get started, but you do not want a topic that is so easy that it is Mickey Mouse, either. You do want a topic that is exciting enough to sustain your interest through the weeks and months ahead.

Browse Through Science Literature and the Internet

The first method is to browse through science books, magazines, and the Internet (especially the World Wide Web). Both your public library and your school library are a treasure trove of good ideas waiting to be uncovered. You could start digging by finding out whether the library has any of the books listed in "Appendix D: Further Reading" at the end of this book. Another way to start is by going directly to the shelves in the library that contain the science books; you will find them placed conveniently together, because the nonfiction books are organized by subject. In the Dewey decimal system, the science books are the 500s, and books on science fair projects in particular have the number 507.

Some ideas will take more work to uncover than others. It would be fairly easy, for example, to pick up a book of experiments and fit your topic around one of the experiments in it. Some of the books have experiments only in a specialized topic like solar energy, electricity, or chemistry; others are more general. Many of my students who are doing science fair projects for the very first time get their ideas from one of these books. Later on, after they have had more experience and have participated in science fairs, they are better able to come up with experiments on their own. For a first-timer, however, books with ready-made experiments are an acceptable place to look.

Another way to uncover an idea at the library is to read a magazine article that interests you and design an experiment to tie in with it. One of my students, Maja Bombaugh, read a five-paragraph article in *Gardens for All* describing the Japanese method of growing apples organically. The Japanese farmers were putting paper lunch bags around the apples to prevent insect damage. Maja decided to find out whether this would work in Ohio, so she designed her own experiment around this idea.

Publications that are particularly good to browse through include *Organic Gardening, Scientific American, Smithsonian, Modern Psychology, Science News,* the science section of *The New York Times, Radio Electronics, Popular Mechanics, Mac-World, PC World, Science Today, National Wildlife, Audubon,* and *National Geographic.* There are others that may interest you, too.

Sometimes a science textbook can be just the source you need. There you may find an experiment that can be adapted creatively to fit your topic. One of my students, David Daly, was interested in space travel. He needed an experiment to tie in with his interest. Since David had no means for conducting a space flight himself, he had to come up with another idea for an experiment. He knew that a system for recycling oxygen and carbon dioxide would be needed for long flights; therefore, when he saw the classic experiment used in many biology classes for demonstrating oxygen production by aquatic plants,

Surfing the Internet and browsing through magazines and books at the library can lead to ideas for possible topics. Do not neglect any of the library's resources. Consult electronic listings of articles or use the *Reader's Guide to Periodical Literature*. Your most valuable resource, of course, is the librarian.

David knew he had found a possible idea for his experiment. By measuring the oxygen output of various plants under various conditions, David could try to find the optimum plant and conditions for space travel. The drawing on page 21 shows the setup for David's experiment.

The Internet, and especially the World Wide Web, is yet another rich source of science fair ideas. Powerful search engines and browsers can help you surf the Net. Using any of the Internet addresses at the end of this book in "Appendix E: Internet Addresses" is an excellent way to get started. Just one word of caution, however: Be sure to write down the particular address where you found items of interest. It is too easy to go from link to link and then never be able to get back there again.

Talk with Other People

The library is one of the best places to start looking for your project idea, but you may need even more help. So where can you go from there? The second method is to talk with other people, especially your parents. You need not think that your project is less your own if you use ideas that other people have shared with you. Scientists talk with other scientists in order to sharpen their own ideas and learn new information; you can help yourself too by talking with others. Parents are especially good people to talk with because they are so interested in you!

Just in case you are hesitating to go to your parents because you are not sure how they could help, here are two short examples of how other parents have helped. Sarah Speicher knew that she wanted to do her project on the subject of pollution, but she could not come up with an idea of exactly what. She brainstormed with her father, an auto worker with the Ford Motor Company, and together they thought of a project on noise pollution. Sarah's father had a union responsibility to monitor the noise pollution that he and his fellow workers were subjected to in the plant. This gave him access to a special meter that measured the amount of sound present. Sarah was able to borrow this meter and measure the amount of noise pollution at various locations and times in her hometown; her prizewinning project proved that her hometown could be described as "quiet."

Another of my students, Aliza Weidenbaum, wanted to do a project in chemistry, but she did not know enough about

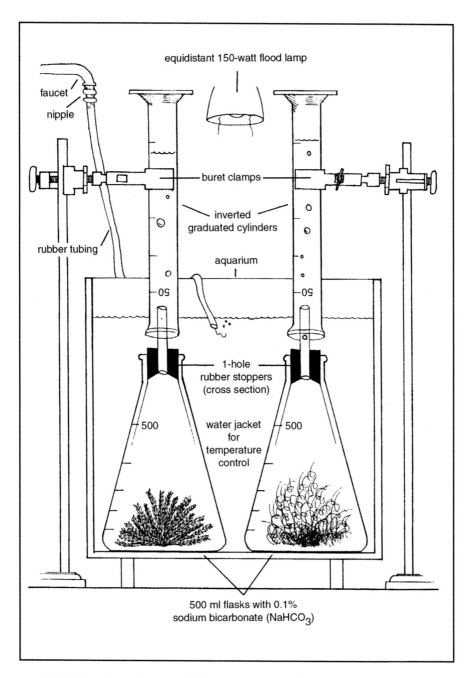

equidistant 150-watt flood lamp

faucet

nipple

buret clamps

inverted
graduated cylinders

rubber tubing

aquarium

50

50

1-hole
rubber stoppers
(cross section)

500

water jacket
for
temperature
control

500

500 ml flasks with 0.1%
sodium bicarbonate (NaHCO$_3$)

David Daly adapted a classic biology experiment to fit his topic of space travel.
Browsing through biology textbooks and talking to his teacher gave him the idea.

the subject to get started. She talked with her father, a former chemist, and he suggested that she work on a food chemistry experiment having to do with tofu. At first, Aliza thought the topic would be too boring. "After all," she thought, "what can be so interesting about a food that my family makes every week by turning soybeans into a curd?" The more Aliza thought about it, however, the better it sounded. She began to realize that although her family made tofu every week, most people had no idea what exactly tofu was, let alone how to make it. Aliza finally decided to make tofu from soybeans and from other legumes as well. When she compared the amount of curd produced from the various legumes, she concluded that the ancient Chinese, who first made tofu, must have known how unique soybeans were, because soybeans produced the most curd.

Both Sarah and Aliza got just the help they needed for a good topic from their parents. However, I do not mean to suggest that parents are the only people you could talk with. Older students who have done science fair projects can draw upon their own experience and those of their friends to help you. They are often eager to share their expertise with young people. When you go to an adult professional, however, remember to be polite so that the next student asking for help will also be welcomed. Be sure to bring your logbook when you talk to the adult, and have a few questions ready in advance.

Follow Your Own Interests or Curiosity

Although people and books are great sources of project ideas, you may be an independent person who wants to come up with your very own topic. The third method is simply to follow your own interests or curiosity.

Aliza used her knowledge of how to make tofu to investigate whether legumes with a higher protein content would yield higher amounts of curds. She got her idea for the project by talking with her father.

Do you have a hobby that could provide you with a good science fair topic? One student, Tim Young, helped his grandfather raise bees and really enjoyed working with them. He decided to do an experiment on the crystallization of honey. His report discussed all aspects of apiculture (beekeeping). Since he already had a strong interest in and background knowledge of the subject, his report was a lot easier to write than it would have been had he written on an unfamiliar subject. Other projects that grew out of students' hobbies include a go-cart that one student built to compare gasoline mileage under different road conditions, the natural dyes a student tested with various color-fixing agents called mordants, a radio built by a student to test the reception of a

Do you have a hobby or special interest that would make a good science fair topic?

local station under different weather conditions and times of the day, a miniature car model that another student built to compare engine performance of various fuels, and a paper airplane experiment in which a student made predictions of and then tested how various wing designs would affect flight distance.

Are you athletically inclined? An interest in physical fitness can be used to devise a whole series of experiments. One student, Tim Russell, measured the pulse rate of his wrestling team after a certain amount of exercise both before the wrestling season started and at the finish. He also compared this measure of cardiovascular fitness with students who were not on any athletic teams. Another student, Ruth Chiego,

Do basketball athletes six weeks into their training have lower pulse rates after exercise than they did before training started? There are many questions that could be turned into science fair projects dealing with athletics.

timed her teammates' lap runs after they had followed carefully controlled diets with and without complex carbohydrates.

Pets are a real natural for science fair projects. One of my students studied her tarantula's behavior before and after molting. Another student bred her feeder guppies into fancy guppies. Other students have investigated food preferences of their pets, the effect of exercise on water consumption, and the effect of some environmental aspect such as color on their animals' behavior. One word of caution, though: Many science fairs have strict rules for using vertebrates (animals with backbones, such as dogs, cats, frogs, fish, and humans). If you do not first check the rules and abide by them, you could be disqualified. These rules are intended to protect animals from cruel and inhumane treatment. Even if the rules did not exist, however, few judges would give high marks for a project that caused an animal to suffer.

Plants are another good source for science fair ideas. There are a multitude of different things you could test with them. One of my students tested the effect of different plant foods on the growth of spider plants. Another tested the effect of microwave radiation on bean leaves, and yet another tested the effect of acid rain on radish plants. One of the most interesting plant projects a student came up with was Phadra Khalil's experiment with hydroponics, the growing of plants in a mineral solution of water. Phadra found an old library book that contained a recipe for a hydroponic solution devised by a chemist in the late 1800s. She used this recipe to grow spider plants and compared their growth to the growth of spider plants in a modern hydroponic solution. She also raised some spider plants in plain water as a control. The old recipe worked best!

In summary, by browsing through science materials at the library, talking to people, and following your own interests, you will find a good project that is right for you. Once you have your topic, you will need to design an experiment that is related to it. This leads to the next chapter, in which you will learn how to narrow your topic down to a good research question and write a proposal for your experiment.

Step 3

Designing Your Experiment

After you have chosen your topic, you will be ready to start the real science part of your project: designing and carrying out an experiment. Most of my students like this part best; it is fun to conduct your very own experiment! This chapter will give you the guidelines you need to do the job well.

Ask a Narrow, Focused Question

An experiment is designed to answer a question. This means that you first have to think of a question to ask. The best kind of question is a narrow, focused one. If the question is too broad or too vague, your experiment will not provide enough data to answer it. For example, suppose you ask the question: How much do birds weigh? Unless the question is reworded, you cannot answer it. It would all depend on the type of birds being weighed. What kind of birds? How old? On what and how often have they been fed? Suppose the question were changed: How much will 50 female mallard ducks weigh three months after hatching if fed all they will eat of commercial duck food? That question would provide a better starting point from which to conduct your experiment and a reasonable chance of finding a meaningful answer.

Should You Use a Control?

Most well-designed experiments require the use of controls. A controlled experiment has at least two setups: One part may be called the test, the other the control. Suppose, for example, that we wanted to know if adding vitamins and minerals to commercial duck food would help mallard ducks grow better. We would need to test two groups of ducks. One group of ducks would be the control group and would be fed just the commercial food. The second group would be the test group and would be fed the commercial food with vitamins and minerals added to their diet. At the end of the experiment, we would compare the two groups to see whether the vitamins and minerals really helped.

Dina Kountoupes, one of my former students, designed a controlled experiment that might help you see how a control is used. Dina wanted to find out if subliminal messages really influence people. A subliminal message is a message that is spoken so softly or flashed on a screen so quickly that you cannot consciously hear or see it, but your subconscious receives it. For example, some department stores whisper the subliminal message, "You will not steal . . . you are an honest person" during broadcasts of Christmas carols. The customers cannot consciously hear the message, but store managers hope that the message will be received subconsciously, discouraging customers from shoplifting. Dina's research question was, "Will a whispered subliminal message cause more people to choose a certain cup under which a pea seed may be hidden than the other four cups?" Her hypothesis was that yes, subliminal messages can influence a person's choice in a game of chance. She thought twice as many people would choose the cup that the subliminal message urged them to pick. She began setting up her experiment by taking five paper cups and numbering them from 1 to 5. Next, she recorded a popular rock song twice. She left the first recording exactly as it was, but on the second recording she dubbed in the whispered subliminal message "Pick number 4."

The same one hundred students were tested with the same five cups, with and without subliminal messages.

Then, she was ready to conduct her experiment. Dina tested her fellow students one by one. She told them that there was a pea seed under one of the cups and that she wanted them to guess which cup it was. Dina played the music without the subliminal message for 100 students and recorded each of their responses. This was the control for her experiment. She found that they picked each of the cups roughly an equal number of times. Dina next tested the same 100 students with the same five cups in the same manner except that this time she played the music with the subliminal message. This time, more than three fourths of the students picked cup 4!

Dina concluded that her hypothesis was probably correct: Subliminal messages can influence the choice a person makes in a game of chance. But what if she had not used the tape without the subliminal message? Could she have reached the same conclusion? Without the use of a control, Dina could not be sure whether three fourths of the students would have picked cup 4 regardless of the subliminal message. By using a control, Dina could compare the number of times cup 4 was picked with and without the subliminal message and, on this basis, reach the conclusion that her hypothesis was correct.

Keep the Conditions the Same— Independent and Dependent Variables

Something else is important in understanding Dina's controlled experiment. Dina used the *same* cups and the *same* students, and she conducted all her trials in the *same* manner. The ideal controlled experiment is one in which all conditions are the same for both the control setup and the test setup except for the one specific thing being tested.

The one specific thing being tested is called the *independent variable* (or sometimes the *manipulated variable*). In Dina's experiment, the subliminal message was the independent variable because it was the one thing that was different between her control setup and her test setup. The subliminal message was *not* present in her control setup but it was present in her test setup. Everything else that might have affected her experiment, such as the size of the cups or the directions she gave, were kept the same. Dina wanted to know the effect of the subliminal message on which cup students would choose. She was not interested in knowing what effect the size of the cup might have or how giving her directions differently each time would affect a student's choice. She kept the conditions the same for both the controlled setup and the experimental setup by watching out for *potential* variables.

Another important term you should know is *dependent variable*. The dependent variable is the one that you measure and keep track of. You are interested in figuring out the effect of the independent variable on the dependent variable. For example, the subliminal message was the independent variable and the choice of which cup students picked was the dependent variable in Dina's experiment (the choice *depended* on the subliminal message).

In actual practice, you would have to be very clever to keep all conditions exactly the same for both the control and test setups; it is almost impossible to do. However, you should strive to come as close as possible to having all conditions the same. Not all well-designed experiments have both a controlled setup and a test setup. All well-designed experiments do, however, try to keep all conditions the same except for the one specific thing, the independent variable, being tested. Examples of scientific work and experiments without a control would be some of the observational studies done on animal behavior, volcanic activity, and weather. Also, surveys of people would not include a control, either.

The More Data You Collect, the More Believable Your Results Are

Steve Lewis, another former student of mine, had a project without a control. He wanted to know whether the brain dominance of a student would influence how he or she would perceive an ambiguous figure. Steve's project was a lot of fun; here are a few terms to help you understand it better. Your brain is divided into two distinct halves (also called hemispheres), the right half and the left half. Some people's behaviors are controlled by the right halves of their brains; they are called right-brain dominant. Other people are controlled by the left parts of their brains; they are called left-brain dominant. Most people are neither strongly right- or strongly left-brain dominant; they are closer to being whole-brain dominant. Steve wanted to first identify those few individuals in the school who were strongly right- or left-brain dominant and then find out whether they saw ambiguous figures differently. (An ambiguous figure is one that can be seen in more than one way. See the illustration on page 34 for examples of ambiguous figures.) Even though Steve had no control, he had to be very careful to survey all the students for brain dominance in the same way and to test the right-brained and left-brained students in the same manner. He accomplished this by administering the same test and giving

These three figures are called ambiguous because you can see them in two distinctly different ways. The middle figure can be seen either as a vase or as two faces. What about the other two?

the same directions each time. He wrote a computer program to handle the large amounts of data, since he surveyed the entire student population of his school.

Steve's project required this large amount of data. Suppose Steve showed an ambiguous figure to only one right-brain-dominant student and one left-brain-dominant student and found no difference in their perceptions. He could not be sure whether that was because there is no difference to be found or whether those particular individuals were not typical of right- and left-brain-dominant students. Now suppose Steve found many right- and left-brain-dominant students and tested them with several ambiguous figures. If this time he found no difference in their perception, he could be more certain that there is no difference to be found. His experiment clearly illustrates this important point: *The more data you collect, the more believable your results are.* (In case you are curious, Steve concluded that brain dominance probably has no effect on the way ambiguous figures are perceived.)

Important Things to Remember About Designing an Experiment

In summary, there are three main guidelines for designing your experiment. First, if at all possible, use a control. The control enables you to make comparisons. Second, keep all conditions the same except for the one thing being tested. If you do not keep everything the same, you will introduce more variables than the one thing you are trying to test. This will make your data worthless. Finally, collect as much data as is reasonably possible. The more data you collect, the more believable your results and conclusion will be.

Collect as much data as is reasonably possible. The more data you collect, the more believable your results and conclusion will be.

Writing Your Proposal

Now that you know the guidelines for designing an experiment, you need a format to write up what you are going to do. A shorter name for "what you are going to do" is *proposal.* Your proposal should include a title for your experiment, the question you are trying to answer, a hypothesis (which is your best educated guess about what the answer will be), a list of the materials needed to perform the experiment, and a list of the procedures.

You need to write very concisely so that your proposal is not more than two thirds or three fourths of a single-spaced typed page. The reason you should keep your proposal this short is so that you can use it later as part of your summary for the experiment. The summary is a very important document that you will learn more about later; you will include one copy of it with your research report, and you will attach another copy of it to your visual backdrop. Right now, though, you just need to learn a good format for the proposal.

Let's use Dina's experiment on the effect of subliminal messages to see what a proposal might look like. Notice how the indenting is done. This is a good format to use for a proposal of any experiment. If you study it carefully and use it as a model for your proposal, you can't go wrong.

Effect of a Subliminal Message on Choices Made in a Game of Chance

Question: Will a whispered subliminal message cause more people to choose a certain cup under which a pea seed may be hidden than the other four cups?

Hypothesis: Twice as many people will choose a certain cup if they receive a subliminal message to do so than if no subliminal message had been given.

Materials: Five identical paper cups, marker, pea seed, two blank audio tapes, recording of a song, recording and playback equipment, logbook, and pen.

Procedures: **A.** Number each cup with the marker from 1 to 5.

B. Record a popular rock song onto two blank tapes and label the tapes A and B.

C. Dub the whispered subliminal message "pick four" onto tape A, but leave tape B as it is.

D. Put the pea seed under one of the cups and line up the cups in numerical order.

E. Test 100 students one at a time by asking them to pick which cup has a pea seed under it while tape B is playing. Test the same 100 students with the same five cups but with tape A playing. Give each person just one pick with each tape.

F. Record each response (choice) carefully in the logbook; be sure to identify which tape was being played when the choice was made.

G. Organize the data into chart or graph form.

Step 4

Keeping Track of Your Experiment

Since your experiment is the most important part of your science fair project, keeping careful records of it is the most crucial task of all. Now is the time to put that logbook to really good use! Remember that the more data you have, the more believable your results will be. You want to collect as much data as is reasonably possible.

Using Measurements and Numbers

Be sure to keep track of your experiment with measurements and numbers rather than subjective observations. Experiments that include measuring are called quantitative, because you are keeping track of the quantity of something. Experiments and scientific works without measurements are called qualitative. You want your

Tracy kept careful quantitative records of her tarantula's behavior before, during, and after molting.

experiment to be as quantitative as possible, because then you have something concrete on which to base your conclusion.

Let's look at an example of what we mean by making the experiment quantitative. Suppose Tracy Arbogast looked for behavioral differences in her tarantula before and after molting. Suppose further that one of the differences she observed (but did not write down) was the spider's eating habits. If Tracy did not keep any records but simply concluded, "The tarantula ate more after molting than before," how could she support her conclusion? If she did not have anything written down, what could she show the judges as proof? Now suppose Tracy did keep careful records of how many crickets (her tarantula's favorite food) her tarantula had eaten during a six-month period. Suppose further that her tarantula normally ate six crickets over a two-week period, did not eat any the whole month before molting, and ate eight the day following molting. With these figures, Tracy has very specific proof for her conclusion and can substantiate her claim with numbers. Notice one other important point in this example: Tracy's conclusion was that the tarantula ate more, rather than, say, that the tarantula felt hungrier. Tracy kept to a conclusion based on measurement and numbers rather than the unmeasurable and subjective field of feelings.

Raw Data

The measurements you make and enter into your logbook are the raw data of your experiment. You can reorganize this raw data later into charts and graphs, but you must always keep this original copy of your raw data. That is why you need to keep it in your logbook and not on scraps of paper or in your head. Also, write with a pen. All data must be recorded in ink. This will remind you that you must never change your raw data; what you record is what really happened, whether or not it is a result that you expected to get. There are no wrong results in a science experiment, only unexpected ones. If, on the other hand, you find that you recorded a measurement that you had measured wrong (you will find this out by remeasuring it immediately after your first measuring), simply cross out the first measurement with a single line, write your second measurement next to it, and label the first one *error*. Do not erase it.

Using the Metric System

Be sure to measure in metric units. Metric units are degrees Celsius instead of Fahrenheit, centimeters instead of inches, grams instead of ounces, and liters instead of quarts. There are other units for any other kind of measurement you might want to take as well.

The beauty of the metric system is that it is based on the number 10. Any American who says that the metric system is too complicated to learn probably does not realize that we use the metric system every day: Our money is based on the metric system, thanks to Thomas Jefferson, who set it up that way. Ten pennies make a dime, just as ten centimeters make a decimeter. Ten dimes make a dollar, just as ten decimeters make a meter. Simple yet elegant! You never have to divide the number of inches by 12 to get the number of feet again; to convert from centimeters to meters, just move a decimal point. For example, 324 centimeters is 3.24 meters. Throughout the whole metric system, all you have to do is move a decimal point. Wonderful, isn't it?

More important than the simplicity of the metric system though, is that scientists the world over use metric. Most judges will deduct points from your project if your measurements are not in metric. Take note also that it is much, much easier to, say, invest in a metric thermometer (if you need to take temperature readings) before you set up your experiment rather than measuring in degrees Fahrenheit and converting them into degrees Celsius after your experiment is done. Converting any of our customary units (such as degrees Fahrenheit) into metric units is awkward, and it increases the chance of error creeping into your work. Besides, you will have enough to do at the end of your experiment without adding one more chore!

Photographing Your Experiment

Another way to keep track of your experiment besides recording measurements in your logbook is to take pictures as you go along. Pictures can help out in a number of ways. Best of all, they can serve as documentation of the progress of your experiment. If, for example, you were keeping track of the effect of acid rain on the germination and growth of radish plants, you could record the appearance of the plants week by week by taking photographs as well as by taking careful measurements and recording notes in your logbook.

Another way photographs are important is to demonstrate your apparatus. Some apparatus or other parts of your experiment might be too large or too hazardous to use as part of your visual display. If your experiment involves disease-causing organisms, microbial cultures, food materials, flames, or certain types of lasers, you probably will not be allowed to display them. Having photographs of these materials in use during your experiment would be a good substitute for displaying them at the fair. You just need to remember to take the pictures before your experiment is over.

Do not be shy about putting yourself in the pictures. Pictures with a human subject (you!) are much more interesting than those without. Also, having yourself in the pictures documents your own personal involvement in your project.

After you are finished collecting your data (both recording your measurements and taking the pictures), you still have two important tasks to do before your experiment is finished.

First, you have to organize your measurements into charts or graphs, and second, you have to come up with a conclusion.

Organizing Your Data into Charts and Graphs

Charts (or tables) are a good way to show a complete set of data in full detail. By organizing all your figures into appropriate columns, you can fit a large amount of information into a reasonably small amount of space. Frequently, however, a graph works better than a chart. Why? Because graphs show overall relationships, which is more important than detailing every little datum you collected. Graphs are designed to reveal relationships between independent variables and dependent variables. In Dina's experiment with subliminal messages, for example (see page 30), the subliminal message was the independent variable and the choice of cups was the dependent variable. Her bar graph had one set of bars to show how often students picked each of the five cups without the subliminal message and another set of bars to show how often students picked each cup with the subliminal message.

Fatima Khalil used tables and a bar graph in her project on iron deficiency anemia. (*Anemia* is a medical term for a lack of red blood cells. Iron deficiency anemia is the form of anemia that occurs when the body does not have enough iron available to make red blood cells.) One of Fatima's research questions asked how many of her fellow seventh-grade students would have high, average, and low levels of red blood cells.

Fatima's hypothesis was that more boys than girls would have higher-than-average levels of red blood cells, and that more girls than boys would have lower levels. In order to find out if her hypothesis was right, Fatima and the school nurse used an instrument for measuring hemoglobin (the substance that makes red blood cells red), a hemoglobinometer. Fatima recorded the name of each student tested and the result of his or her test in the logbook using the units of milligrams of hemoglobin per deciliter of blood. When Fatima and the nurse had finished testing all the students who could be persuaded to give a blood sample, Fatima had a long, two-column chart that included all the data she had taken.

Fatima put the raw data on a spreadsheet, then rearranged them by using the Sort command to separate the test results of the boys from those of the girls. This arrangement was more helpful, but even so, Fatima could not tell just by looking at it whether her hypothesis was right or wrong. It gave the same visual impression as a page from a telephone book! She decided to use the Sort command again, rearranging the data from the highest levels to the lowest levels. Then it was easy for her to count the numbers of boys and the numbers of girls who had high, average, and low levels of hemoglobin. Since a lot more girls were willing to be tested than boys, however, she could not directly compare the numbers of boys with low, average, and high levels with the number of girls with low, average, and high levels, so she expressed the numbers as percentages and then made a bar graph.

Bar Graphs, Pie Charts, and Line Graphs

Fatima's bar graph clearly demonstrated that her hypothesis was correct: Of the students she tested, more boys had

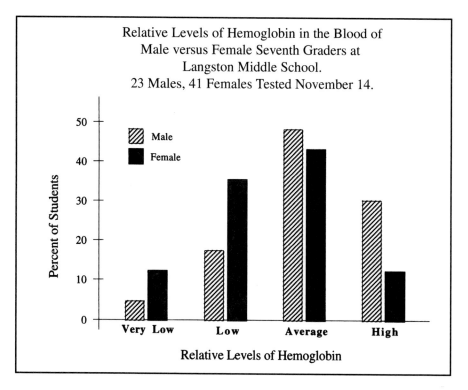

Relative Levels of Hemoglobin in the Blood of
Male versus Female Seventh Graders at
Langston Middle School.
23 Males, 41 Females Tested November 14.

Graphs show relationships with one quick look. Fatima's bar graph shows the relationship between students' gender and their hemoglobin levels.

higher-than-average levels of hemoglobin (and, therefore, presumably more red blood cells); and more girls had lower-than-average levels. Her bar graph showed the relationship of gender, her independent variable, to levels of hemoglobin, her dependent variable, much more clearly than her spreadsheet of all her collected data.

Bar graphs are especially effective for showing relationships, but they are not the only way of displaying your data. Two other types of graphs that are commonly used are the pie chart and the line graph. A pie chart is especially good for showing results of surveys. A line graph is often the most appropriate choice when you are plotting any measurements versus time. Be sure if you are making a line graph that you put time, the independent variable, on the horizontal axis (also called the x-axis).

Labeling Your Graphs

Whatever type of graph you use, be sure to label it. A graph without labels is meaningless. There are at least three places where you need to label a bar or line graph: (1) at the top with a title, (2) on the bottom for the horizontal axis, and (3) on the left side for the vertical axis. Be sure you tell not only what the bars or lines represent but also the units. If you have multiple sets of data on your graph, you will need to use a key as well as the labels to identify one set from another. For example, Fatima made her bars representing the data for boys striped and the bars representing the data for girls solid. Then she made a key that identified the striped bars as data on boys and the solid bars as data on girls. For pie charts, be sure to include a title, and label each piece of the pie, noting the percentage of the pie each piece represents.

Drawing a Conclusion

The second important task comes after looking carefully at your charts and graphs. You should be able to come to a conclusion concerning your hypothesis. A conclusion is just one or two sentences saying whether your hypothesis was right or wrong.

Iron Deficiency Anemia in Early Adolescents

Question: What percentage of seventh-grade males and females have low levels of iron in their blood? Do these students eat less iron-rich foods than students with high levels of iron in their blood?

Hypothesis: Less than half the male students and over half the female students have low and borderline low levels of iron in their blood; these students (on the average) eat less iron-rich foods than do students with high levels of iron.

Materials: Parental consent forms for blood tests, seven-day diet sheets, sterile lancets, cotton, hemoglobinometer, logbook and pen.

Procedures:
1. Petition the school ethics committee to work with human subjects. Do not proceed until permission is granted.
2. Give each student volunteer a parental consent form with instructions. Collect them the next day.
3. Give each student a food diary sheet with directions for each of seven consecutive days. Urge them to follow through in completing them.
4. Collect completed diary sheets. Calculate numbers of iron-rich foods consumed weekly by each volunteer.
5. Be present while the school nurse, an R.N., administers the blood tests and record data for her.
6. Analyze data from the blood tests. Count the number of male vs. female students who have low, average, and high iron content by their hemoglobin count.
7. Compare the average, weekly intake of iron-rich foods consumed by students with high versus low iron content in their blood.
8. Organize the data in graph form.

Results:

Relative Iron Levels in Blood of
Male versus Female Fellow Seventh Graders

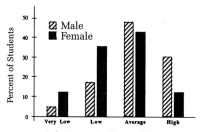

Relative Levels of Hemoglobin

Average Number of Iron-Rich Meals:
Iron-Rich versus Iron-poor Blood

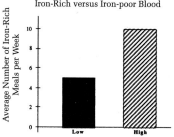

Relative Levels of Iron

Conclusion: The hypothesis is correct. More females than males have low levels of iron in their blood; also, a positive correlation exists between the amount of iron-rich foods eaten and levels of iron in the blood.

Notice how Fatima's summary sheet fits within a typewritten page. Also take careful note of the indenting and where lines are skipped.

From Proposal to Summary

Now that you have finished your experiment, you can add the *results* and conclusion to your written proposal and turn it into a summary. Skip a line after the last procedure that you wrote on your proposal and write *Results* on the left-hand side under the labelings for the other sections (Question, Hypothesis, Materials, and Procedures). Put a small copy of the most important graph(s) in this section. Skip another line and write *Conclusion*. Put your one- or two-sentence conclusion in this section. Your whole summary should fit on a single-spaced typed page. Many district and state fairs require that each project have a summary for the experiment and that the summary be no more than one typed page long.

You will put one copy of your summary in your research report and another copy on your display board as explained later. Your summary is the nucleus of your science fair project.

Step 5

Writing Your Research Report

All science fair projects need a research report. There is no single correct format for your report unless a competition you are going to be in specifies one. If you do not have to meet such a requirement, the specific directions given in this chapter will give you a general-purpose format that can be used for any type of investigative project you are doing.

The best way to go about writing your paper is not the most obvious. You should start from the back. That is, you should write the last section of your report first, then proceed to the second from the last, and so on. This reverse order may seem strange, but it will give you the time you need to finish collecting data and come to a conclusion in your experiment. You will not be able to write the first sections of your report until your experiment has been completed, whereas writing the last sections of your report first can actually help you perform the experiment.

Bibliography

The bibliography goes at the very end of your report, so it should be the first section that you write. It is a list of all encyclopedia articles, books, magazine and journal articles, pamphlets, World Wide Web material, and other sources that you are using to write your research report. Since you need to gather these materials before you begin your report, this is the best time to list them in a bibliography. Listing them now eliminates trying to relocate returned library materials after you finish writing the other sections of your report because you forgot to jot down bibliographical information when you first started. If you need additional sources later on as you are working on your project, you can always insert additional entries into your bibliography as needed.

When you are gathering your sources, try to select a good variety. Using just one or two encyclopedia articles or getting all your information from a single book is not sufficient. The Internet abounds in fascinating avenues for learning; some good addresses to get you started can be found in "Appendix E: Internet Addresses." Be sure that you actually take the time to *read thoroughly* some of the information that you find on the Internet, however, and document the location (i.e., Internet address), the date it was published, and date that you obtained it. (Note: You also need to be concerned that the information you pick up from the Internet is *reliable.* All of the addresses in Appendix E have a good reputation for being reliable.)

The latest information on your topic will also be contained in current magazine and journal articles. Finding these articles, however, can be like looking for a needle in a haystack. Your local library may have computerized listings of articles or special databases organized by subject and author, which can help you. If your library does not have the computerized listing you need, it undoubtedly will have the *Readers' Guide to Periodical Literature* or other special indices, which are also excellent resources. As always, the reference librarian is your greatest resource of all. Librarians will enjoy helping you, so be sure to ask for some guidance in locating good sources.

Gather at least five sources for your report. When you list these alphabetically in your bibliography, you must provide certain information in a particular order and with particular punctuation. Some district and state science fairs specify that a certain style, for example the APA, which is the American Psychological Association style, or the MLA, which is the Modern Language Association style, be used for writing a bibliography. All styles require nearly the same information, but the particular order and punctuation differs. For books, you will need to give the author's name, title of the book, city where it was published, name of the publisher, and the year it was published. There should be a manual explaining the style in your local library. Since the APA style and MLA style (and other styles as well) change from time to time, it would be good to check the form you are using with one of your teachers or the librarian.

"Appendix D: Further Reading," at the end of this book can help you see how bibliographic entries look in a style used by many publishers.

Background Information

The background information part of a research report is usually the longest section. Anywhere from five to fifteen double-spaced typed pages is a good length. It presents general information on the topic but no specific information on the experiment. If, for example, you were doing an experiment on the food preferences of starfish, your background information might include a description of the life cycle of starfish, different kinds of starfish, how starfish are classified by biologists, special adaptations (both behavioral and structural) of starfish, and information on the range and habitats of starfish. You also might include some of this same type of information on the organisms that starfish eat.

In the process of reading, taking notes, and writing the background information for your report, you will gain a broader understanding of your topic and be better prepared to answer any questions the judges may have.

Further Explanation of the Experiment

Your summary sheet tells what is most important to know about your experiment, but this section of your report, the further explanation of the experiment, will provide you with an opportunity to elaborate. You should highlight any difficulties that came up and how you overcame them. You should also include a discussion of further research that the results of your experiment might suggest to you. (There are always interesting tangents, or related thoughts, that keep presenting themselves when an experiment is being done.) Some judges will ask you what further research is suggested by your experiment. Any other pertinent and interesting information concerning your experiment could also be included in this section.

Summary Sheet

The summary sheet consists of the proposal for your experiment plus the results and a conclusion (see "Step 4: Keeping Track of Your Experiment"). If you set up and performed your experiment any differently from what you had anticipated and written in your proposal, you should rewrite your procedures so that they are accurate. Make sure your procedures are written clearly enough for another person to be able to follow them and repeat your exact experiment. This is very important in science. A good experiment must be verifiable. That is, any other person should be able to repeat it in the same manner and get the same (or similar) results that the first experimenter got.

Abstract

The smallest section of your report, the abstract, may be the biggest challenge of all. In it, you need to tell what is most important in your project in a very limited space. A typical abstract may be as short as one paragraph or as long as one typed page (roughly 250 to 300 words).

Some district fairs specify a length such as fifty words or less. (Be sure to check with your teacher to find out whether the competitions you will participate in have such a limit for the abstract.) If you are trying to keep your abstract this short, you should include a one-sentence description of your experiment, another sentence to highlight the most important results of your experiment, and a concluding sentence (if you have space) that tells what topics are covered in the background information section of your research report.

Table of Contents

Your table of contents will list the sections for your report in the reverse order that they have been presented here: abstract, summary sheet, further explanation of the experiment, background information, and bibliography. Since you will need to know how long each section is, you will write your table of contents last.

Title Page

A good title can be an attention grabber. The time you spend thinking of a good title is time well spent. The most descriptive title would include both the independent and the dependent variables.

Examples for some of the projects already mentioned in this book might be, "How a Tarantula's Behavior Changes Before and After Molting" or "Perception of Ambiguous Figures by Right-Brain-Dominant versus Left-Brain-Dominant Individuals."

Your title should be centered and appear one third the way to halfway down on the page. Your name, the name of your school, and the date should be centered about three-quarters down the page. (A word processor will do the centering automatically if you use the correct command.)

Proofreading

Once you have finished writing all the above sections, you will have a rough draft. In order to really benefit from all the hard work that you have put into it, you need to polish your draft

EFFECT OF TEMPERATURE ON THE GERMINATION OF ARTICHOKE SEEDS

Paula Wightman
Langston Middle School Science Fair
January 28, 2002

Not all title pages need to look exactly like this one. This is a very simple style.

by eliminating unnecessary words, making your sentences simpler, and organizing your paragraphs more clearly. Even though this may seem like a tedious chore, it is important that your writing be as clear as possible.

It is easy for an author to overlook or miss his or her own grammar, spelling, and punctuation mistakes, so be sure to ask someone else to proofread your work also. Your English teacher would be the ideal person to check your writing mechanics, and your science teacher could check for any inaccuracies in your content.

The word processor of a computer is a wonderful tool for revising written work. It can save you hours and hours of time. If at all possible—if a computer is available at school or at home—use it to write your rough draft, and revisions will be a snap.

Step 6

Making an Attractive Visual Display

Catch the judges' eyes with your visual display! The visual display is part of the showmanship of your project. An effective display will draw in other viewers as well as the judges: You could find yourself the center attraction.

Even though a well-done display board and accessories cannot substitute for reliable data and a carefully written research report, they can enhance them. Your visual display should act like an advertisement. It should demonstrate how hard you worked on your project. It should also help you communicate to the judges and other viewers what your project is about and how you went about doing it.

Making your visual display can be one of the most enjoyable and creative parts of your project. This chapter will first give you tips on the overall design of your display board. Specific step-by-step information for constructing your display board and doing the lettering follow.

Keep It Simple

Simplicity is a key word for making an effective visual display. If your display board is too full, people will turn away from it rather than invest the time to figure it out. Aim for your visual display to "tell a story" about what your project is about in just the few moments it would take for a person to read it as he or she slowly walks by. This means you cannot clutter it with all the details of your project; those people who are interested enough to want to find out the details will read your research report.

One simple design for a visual display that works very well is to start with three panels for your display board. In the center panel, put the title of your project on top and mount a copy of the summary of your experiment underneath it. This focuses the viewer's attention, front and center, on the most important information concerning your project: the topic and the experiment. If you have access to a computer, enlarge the font of your summary sheet by 50 percent for better visibility. Next, use the left-hand panel to mount extra graphs and charts of your data. You could even use the same graphs or charts that are included in the results section of your summary sheet but in an enlarged and colored form. Finally, use the right-hand panel to mount photographs of yourself and your experiment. Put the pictures in an organized sequence that tells a story about what you did and how you went about doing it. Pictures are very attractive to viewers; people will almost always come closer to a project in order to look at the

An effective display will draw in other viewers as well as the judges: You could find yourself the center of attraction! One of my students, Collyn Rybarcyzk, attracted the attention of a reporter and found her picture on the front page of the newspaper. Her display included a large overstuffed hamburger and jars of lard for props. One of her backdrop panels had pictures of her experiment.

pictures more carefully. Use short captions with your pictures to explain the story further.

Building the Display Board

Three panels, as described earlier, work well for a display board; however, there are lots of other possibilities that could be equally effective. Some students make an effective use of just two panels, and some use four. One of my students used a large cube and mounted her materials on four sides. Another student made a three-sided pyramid and set it on a rotating base. You could choose to make your display board in any of these shapes, or you could come up with an idea of your own. There are three limiting factors, however, in deciding what shape to use. First, you are limited by size. Most science fairs will give you table space that is about thirty inches wide and twenty-five inches deep, but not necessarily any larger. Usually there is no limit on how high your display board is. (On the other hand, it must fit through the door!) Please check your local science fair rules booklet for details.

The other two limiting factors to keep in mind are that your display board must be portable and it must stand up by itself. This means that you probably do not want your board

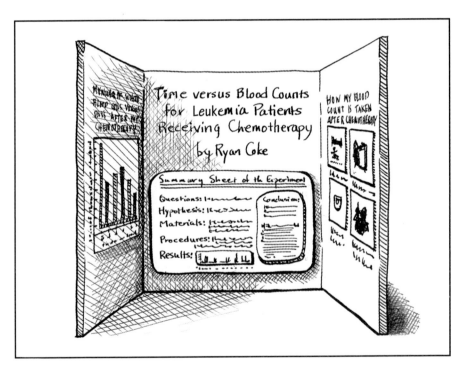

A simple but effective three-panel backdrop could have the title and summary sheet on the center panel and graphs or pictures on the side panels. A student with leukemia did a project similar to the backdrop drawn here.

too tall; making it both portable and able to stand on its own can be tricky enough even with a short one.

Many students make their panels of colored tagboard, which is available in discount stores and art supply stores. You can bind the tagboard panels together with strong tape. The tape will strengthen the panels so that they will stand up on their own while still enabling you to fold them over one another when you need to transport your project. Tagboard has the advantage of coming with a color already on it. You could, however, use another material and add the color yourself. Plywood, corkboard, and cardboard would all be usable materials. You could color them by spray painting them, putting plain or patterned contact paper on them, or even using burlap or another fabric to cover them. Since plywood and corkboard are much heavier than tagboard, you would have to use hinges instead of tape to hold the panels together.

Another method that works even better than hinges is to put eyelets into the wood and drop a bolt, secured with a nut, through them when you are standing up your panels (see the figure on page 62). The bolt and nut are easy to remove in order to separate the panels when you need to carry them.

If a display board made entirely of wood seems too heavy for you but you still want something stronger or larger than tagboard for your panels, you could make a frame out of wood (2" x 2" pine works nicely) and then attach colored tagboard or fabric to it with thumbtacks.

When you are choosing the colors for your display board, remember to keep it simple. Many students using three panels will choose one color for the right and left panels and a contrasting color for the center panel. Then they will use a

Eyelets are a good alternative to hinges for putting the panels together.

third color as a border when mounting their charts, pictures, summary, and so on. A simple three-color scheme can be very effective. Too many colors, though, can make your display look busy and will lessen its attractiveness.

Lettering

Once you have constructed your panels, you are ready to do the lettering. If you have never done any lettering before, you will be amazed at how long it takes! Even if you buy handsome ready-made letters in an art supply store, it will take you time to carefully measure the spaces for them on your panel and attach them. You absolutely must measure where you are going to put your letters so that they will be carefully spaced.

If you have access to a computer, you could choose to word-process your lettering. Some computer software enables you to make banners in a large range of sizes. Using the computer would assure you of neatness and even spacing between the letters. You could mount these banners on a contrasting color of construction paper or border them with a design when mounting them on your display board.

If you do not buy your letters ready-made or type them on a computer, you will have to do your lettering by hand. Although this method is slower, it offers you greater versatility. You can hand-letter using paints, felt-tipped markers, inks, or colored pencils. You should probably use a regular pencil to lightly sketch your letters first, though, before you use one of the other utensils to color them in. Be sure to measure your work; you need to be careful when spacing your letters. You also need to measure when centering your letters or allowing generous borders for them. You can use either all uppercase or uppercase and lowercase letters, but be consistent.

Some students prefer to use a stencil rather than drawing their letters freehand. You could even stencil or trace letters on construction paper and cut them out to mount on your display board. Letters made of construction paper have very even coloring; this makes them appear especially neat.

Neatness Counts!

Neatness counts on your visual display. Judges have a hard time believing that a student has done his or her best job on a project if the display is sloppy and obviously hastily done. Above all, be sure to spell correctly on your display board. A single misspelling in your research report may escape detection, but a misspelling on your display board, especially in a title or large heading, makes a glaring (and potentially embarrassing) error. As soon as you have sketched in your lettering, have a friend or parent proofread it before you go any further.

Finishing Your Visual Display

When your lettering is completed, you are ready to mount the rest of your material. Usually your pictures and graphs will look best if you first mount them on construction paper of a

Any equipment that you used in your experiment can be a prop for your visual display. One of my students built a homemade wind tunnel to collect his data. When he brought it in, other students crowded around to see how it worked.

contrasting color and then put them on your display board. The construction paper will give your material a colorful border that is pleasing to the eye.

Your whole visual display consists of more than just your display board. Three other things that you should have on hand are your research paper, your abstract, and your logbook. Your research paper and abstract should be on the table in plain view. Some students even mount the research paper in a pocket on the display board. The logbook does not have to be in plain view, but it should be available in case a judge asks to see it. If you would like to display it, you may.

The final part of your visual display is the equipment and products of your experiment. These add high interest to your display. They also can be useful when you are explaining your experiment to the judges and other people around you.

Making your visual display can be one of the most creative and enjoyable parts of your project if you make a good effort and can complete it in sufficient time. Select the media that are best for you, remember that neatness counts, and keep it simple, without clutter. If you follow the guidelines in this chapter, you should be able to create a visual display that will enhance, not distract, the viewer's eye and will guide it to the science project itself. Give it your best!

Step 7

Talking to the Judges

The weeks and months of hard work you have spent doing your experiment, research report, and visual display deserve to be presented with the best oral presentation you can possibly give. Your oral presentation can be your best only if you set aside the time necessary to prepare and practice it.

What to Include in Your Oral Presentation

Your speech should consist of three parts: an introduction, some highlights from the background information part of your research report, and an explanation of your experiment.

Start your introduction by telling your name, grade, school, and topic. Next, thank the judges for coming to evaluate your project; after all, they are volunteering their time. Finally, explain to the judges that you are first going to tell them some background information on your topic and then explain your experiment. This lets the judges know what they can expect to hear.

Be sure to stand beside your project when you are giving your introduction and other parts of your speech. You do not want to block your project from their view. Also, be sure to look at the judges. Eye contact is very important.

When you are choosing what parts of your background information to include in your speech, include good basic information. Do not assume that the judges know all about your topic; nobody is an expert on every subject. Ideally, a doctor would be judging the health projects and a geologist would be judging the earth science projects, but sometimes

Be sure to stand beside, not in front of, your project when you are giving your oral presentation. You do not want to block the judges' view.

the reverse happens. So start with the basics before you get technical.

Your summary sheet is a good outline to use for explaining your experiment. Use the pictures on your display board and the apparatus you set up in front to further explain your procedures. Point out the appropriate charts and graphs when discussing your results.

At the end of your speech, invite the judges to read your research report and to ask you any questions that they may have.

Important Dos and Don'ts

Since the judges' time is limited, you should keep your speech within three to eight minutes. This means that you need to be very selective in what you choose to include in your oral presentation. Knowing that you have a time limit may cause you to want to talk fast. Do not! The primary goal of your oral presentation is to help the judges better understand what your project is about. If you talk fast, the chances of another person understanding what you are saying will be greatly diminished.

You cannot read and make eye contact with the judges at the same time. Using note cards is acceptable if you only glance down at them from time to time, but do not read. On the other hand, do not memorize, either. If you are dependent on a memorized presentation, you may get confused if a judge interrupts you to ask a question, or you may panic if you suddenly lose your place. Memorized speeches can sometimes sound too mechanical.

Your personal hygiene and choice of clothes will be noticed by the judges. Even though your appearance should not have any bearing on the evaluation of your project, in the "real world" it could. Poor personal hygiene, sloppy clothes, and chewing big wads of gum can only detract from, not enhance, your project.

Practice Your Oral Presentation

Believe me, you will not be able to put together an effective speech on the day of the fair while you are setting up your display and waiting for the judges! When the judges do finally appear at your station, the only thing that may spare you from being utterly tongue-tied is the knowledge that you have a prepared speech that you have practiced at home and are

Your personal hygiene and choice of clothes will be noticed by the judges. So will any peculiar body movements. Sloppy clothes, wads of gum, and strange gyrations can only detract from, not enhance, your oral presentation.

confident in giving. There are two effective ways to practice your oral presentation, and you should do both: Practice in front of a mirror, and practice in front of people.

It may seem silly to you to practice in front of a mirror, but this is the most effective way to practice eye contact. You need to look at yourself as you give your speech, not only to practice eye contact but also to spot quirks such as swaying from side to side or flailing your arms. Time yourself and make sure that you can give your talk in eight minutes or less.

You should then practice your oral presentation in front of as many people as are willing to listen to you. A live audience will help you find out if your presentation can be understood by others. You could also ask a friend to pretend that he or she is a judge so that you can practice shaking hands and thanking the judge for coming.

Remember the adage "Practice makes perfect!"

What Will Be Judged

Knowing what the judges will be looking for can help you be better prepared when the judging starts. Although no two science fairs are conducted in exactly the same way, almost all science fairs use a combination of the following criteria when evaluating projects: scientific method, clarity, knowledge achieved, thoroughness, and originality.

When the judges are evaluating you on scientific method, they are looking to see if your experiment was well designed and well executed. They will first check your research question and make sure that you had a hypothesis. Next, they will study your procedures to decide whether you were successful in keeping all conditions the same except for your independent variable. Finally, they will carefully examine your results to see whether your data justifies your conclusion.

The clarity of your project depends on how well you write your research paper, how clearly you can explain your project orally, and how well your visual display tells a story. It is important in science that any scientific work can be shared with others. Usually, the points that can be earned for scientific method and clarity are at least as great as any of the other criteria.

Your knowledge achieved is demonstrated by how well you can answer the judges' questions, by how comprehensive your research report is, and by your use of any special techniques in your experiment. Do not feel, however, that you should not admit you do not know something if a judge asks a question you are not prepared to answer. A simple "I do not know the answer to that question but would like to find out" is much better than trying to talk around a question. Judges are not idiots; they will know if you are trying to fool them.

Thoroughness is a measure of how well you followed through on the work involved on your project. Were you persistent enough to collect all the data that was reasonably possible? Did you gather enough sources at the library to write an interesting and informative paper? When obstacles popped up in your experiment, were you able to hang on and work around them?

Your project can earn points for originality in several ways. If the judges have not seen an experiment like yours before, you are sure to earn points in this category. You also can earn points for equipment that you yourself devised or equipment that you used in a unique way to solve a problem. A careful analysis of your data may also earn you originality points. For example, you could find another interpretation of your data besides the obvious one or suggest how to design further experiments to answer questions that your data raised.

Be Proud of Yourself

Winning an award is not everything, but it sure feels nice! What if you do not win an award? There is an element of risk in any worthy human endeavor.

If you have truly worked hard and tried your best, focus your attention on all that you have accomplished whether or not you receive recognition from the judges. You chose a topic, designed and carried out your very own experiment, wrote a

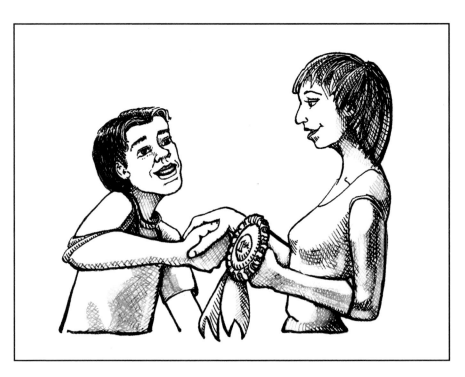

There is more to be gained from science fair projects than awards, but recognition is nice.

research report, built an exhibit to display your work, and made an oral presentation to the judges. That is a lot to be proud of! So whatever the outcome, take pride in what you accomplished; the hard work that you did and the knowledge that you gained are more important than any award or honor. Whether or not you win an award, use the judges' comments and what you can learn from the projects of other students as a means of being a stronger contestant next year.

Once again, good luck to you. I am hoping that, by carefully following each of the steps in this book, you will succeed in completing a project you can be proud of and will receive the recognition you so richly deserve. You will have proved, once again, that young people are capable of making significant scientific inquiry.

Appendix A

Prizewinning Science Fair Projects

BOTANY

EFFECT OF:	ON:
1. Fertilizer such as plant food sold by florists and commercial fertilizers for gardens versus organic material.	Growth of violets in the house and squash plants in the garden.
2. Placing a house plant in an east, west, north, or south window.	Growth and general health of violets.
3. Rooting stimulants.	Growth and development of roots from cuttings of impatiens and begonias.
4. Plant hormone (such as growth stimulators).	Growth and development of bean plants.
5. Lack of specific minerals such as calcium, zinc, or magnesium.	Growth and development of sunflower plants.
6. Companion planting (putting two kinds of plants close together in the garden).	Growth and development of borage with zucchini, parsley with asparagus, and marigold with bean.
7. Different solutions of minerals.	Growth and development of spider plants.
8. Microwave radiation.	Growth and development of bean plants.
9. Various colors of light.	Movement of plants toward the light (phototropism).
10. Age (one, two, and three-years-old) and type of seed (cherry tomato, zucchini, marigold).	Percentage of seed germination.

CHEMISTRY

EFFECT OF:	ON:
1. Sealed-jar method versus evaporation method.	Growth of crystals from several different supersaturated solutions.
2. Protein content of various legumes (including soybean).	Amount of bean curd produced using the standard method for making tofu.
3. Extract from various flowers (boiled water and petals).	Color changes at different pH's.
4. Making candles of paraffin, beeswax, or tallow with or without stearin.	Length of burning, amount of smoke, and scent.
5. Various brands of capsules and tablets.	Length of time to dissolve in weak solution of hydrochloric acid.
6. Various brands of toothpaste and baking soda.	pH level of saliva.
7. Various brands of various kinds of cheese.	Amount of calories as measured in a homemade calorimeter.
8. Different brands and types of vinegar.	Color outcomes of natural dyes (red cabbage, annatto seed, brown onion skin, copper sulfate, madder root, cochineal insects) on blown eggs.
9. Colored filters.	Fading of various colors in various materials.
10. Various laundry detergents.	Ability to clean various types of stains in various fabrics.

EARTH AND SPACE SCIENCE

CORRELATION OR EFFECT OF:	TO OR ON:
1. Depth and location.	Soil pH.
2. Rate of cooling of supersaturated sugar solution.	Size of crystals (to tie in with rock formation).
3. Position of the earth (location in yearly orbit).	Amount of particulate matter containing iron (which could be from comets).
4. Relative humidity and barometric readings.	Weather changes.
5. Wind direction and speed.	Wave size along southern shore of western Lake Erie.
6. Various soils and soil combinations.	Ability to retain water.
7. Various soils and soil combinations.	Rate of heating up when dry and when wet.
8. Various shade trees and ground cover.	Temperature at various heights from the ground (microtemperatures).
9. Wind direction and wind patterns.	Height of water on southern shore of western Lake Erie.

ENGINEERING

EFFECT OF:	ON:
1. Wing design.	Flight distance of paper airplanes.
2. Road surface conditions.	Gas mileage of homemade go-cart.
3. Grease versus synthetic petroleum lubricant.	Energy loss in a system of gears.
4. Different metals as anions of Bunsen battery.	Sustained current output.
5. Various brands of smoke detectors.	Quickness of response to smoke.
6. Various homemade cooling systems.	Inside temperatures of beehive boxes.
7. Various patterns of dimples on golf balls.	Length of trajectory.
8. Various types of G-gauge locomotives.	Amount of draw power.
9. Various homemade, computerized designs.	Ability of "robo-bug" to mimic insect behavior.
10. Various brands of insulation with different R values.	Ability to muffle the sound of a 12-volt, 110-decibel automobile back-up alarm.
11. Various designs of homemade solar ovens.	Length of time to cook a hot dog.
12. Various designs of homemade solar separators.	Length of time to separate salt from water.
13. Various thicknesses and types of wires and other synthetic and natural materials.	Brightness and length of time it glows when put between three D-cell batteries (testing for possible new homemade lightbulb filaments).

ENVIRONMENTAL SCIENCE

CORRELATION OR EFFECT OF:	TO OR ON:
1. Presence of calcium chloride when burning coal.	Output of sulfur dioxide.
2. Water of various pH's (especially at the pH of acid rain).	Growth of radishes.
3. Water of various concentrations of detergents.	Growth of radishes.
4. Location and time.	Amount of noise pollution (as measured with a decibel meter).
5. Time, relative to last precipitation, and location.	Amount of solid particulates in the air.
6. Depth of particular bodies of water.	Turbidity (as measured with a Secchi disk).
7. Can-ban law versus no can-ban law.	Amount of litter made up of cans.
8. Coal ash versus wood ash.	Growth and development of beans, lettuce, and radishes.
9. Various types of kitchen waste.	Length of time to compost.
10. Day of the week, location in the school calendar, grade level and subject, awareness campaign.	Amount of paper waste.
11. Various types of grass and other ground covers.	Ability to capture water runoff on thirty-degree slope.
12. Various types of moist garbage (meat, fish, fruit, bread, vegetables).	Incubation time, number, and kinds of maggots.

HEALTH AND MEDICINE

CORRELATION OR EFFECT OF:	TO OR ON:
1. Inclusion of iron-rich foods versus lack of iron-rich foods in regular diet.	Hemoglobin levels in blood.
2. Gender.	Presence of risk factors of coronary disease in early adolescence.
3. Smoking versus nonsmoking.	Vital capacity of young adolescents
4. Participating in an athletic training program.	Vital capacity and pulse rate after vigorous exercise as compared to resting pulse.
5. Habit of listening to loud music.	Hearing loss.
6. Turning in a swivel chair.	Ability to relocate a pencil dot with eyes shut.
7. Gender, age, and sightedness (amount of nearsightedness, farsightedness).	Peripheral vision.
8. Gender and age.	Adolescent growth.
9. Age, gender, brain dominance.	Perception of five optical illusions.
10. Age.	Performance of NBA basketball players (as measured by five years of statistics).
11. Fat-free versus non–fat-free food.	Food preferences for a healthy diet.
12. Various brands of gum after two minutes of chewing.	Length of time to disappear in a test-tube of pepsin and hydrochloric acid (artificial stomach).
13. Various brands of a variety of fruit juices.	Amount of vitamin C.
14. Various kinds of music.	Pulse rate while resting.
15. Various scents such as peppermint, vanilla, banana, strawberry, lemon, and coconut.	Ability to memorize the scent and identify it again.
16. Length of time after chemotherapy and stage of leukemia.	Blood cell count.
17. Asthma.	Vital capacity of adolescents.
18. Age.	Maturity of male voice as measured by amount of harmonics.

MICROBIOLOGY

EFFECT OF:	ON:
1. Various kinds of mouthwash and toothpaste.	Growth of bacteria.
2. Temperature.	Growth of *Escherichia coli*.
3. Moisture.	Growth of mold.
4. Various brands of bread.	Growth of bread mold.
5. Various brands of detergents.	DNA transfer to bacteria.
6. Various types of pure cultures of cells.	Amount of DNA collected using gel electrophoresis.
7. Various antibiotics.	Bacterial growth on broccoli and turkey.
8. Inoculation of nitrogen-fixing bacteria.	Germination and initial growth of snap peas.

PHYSICS

EFFECT OF:	ON:
1. Size of circular opening.	Height of water column (Bernoulli effect).
2. Size of violin.	Purity of sound of a string.
3. Angle of takeoff, amount and kind of fuel, type of engine.	Trajectory of model rocket.
4. Temperature.	Strength of magnets.
5. Length, width, and material of tube.	Pitch at which it resonates.
6. Temperature.	Specific gravity of certain liquids (useful for constructing a Galileo thermometer).
7. Different powders (parsley, sage, flour, Day-Glo powders).	Patterns produced on negatively and positively charged surfaces (Lichtenberg figures).
8. Use of various liquids such as juices and solutions as electrolytes.	Amount of current conducted through a DC circuit.

PSYCHOLOGY

CORRELATION OR EFFECT OF:	TO OR ON:
1. Sleep deprivation.	Hemoglobin levels in blood.
2. Learning style of students and learning style of teacher.	Ability of student to succeed in a particular class.
3. Making pauses when orally giving a list of words.	Short-term memory recall.
4. Names of colors written in contrasting color.	Length of time needed to read them (Stroop effect).
5. Motivational factors such as self-esteem and life goals.	School success as measured by GPA.
6. Self-professed credence in astrology.	How well false horoscopes matched with person's interpretation of events in personal life.
7. Self-reported previous diving experiences.	Amount of lead weight the diver uses.
8. Handedness.	Syncreatic vision (i.e., the ability to clearly sight foreground and figure at the same time).
9. Self-reported childhood experiences.	Water preference as adult and adolescent (city, well, filtered, sulfur, and spring).
10. Place of residence (urban, suburban, rural).	Ability to recognize everyday sounds.
11. Volume and type of music.	Reading comprehension.
12. Handedness.	Learning style.
13. Playing on home field.	Score outcomes in NBA games.
14. Age and gender.	Ability to inhibit blinking.
15. Age and gender.	Type of body language used when talking to same-aged person of opposite sex.
16. Number of people in elevator.	Voluntary formation of people in elevator.
17. Changes in barometric pressure.	Amount of disciplinary actions given by teachers and school administrators.

ZOOLOGY

EFFECT OF:	ON:
1. Temperature.	Growth and development of tadpoles.
2. Exercise.	Amount of water that horses drink.
3. Whole wheat, soyflour-enriched bread.	Growth and development of mice.
4. Color of environment.	Feeding behavior (frequency and amount) of mice.
5. Intestinal parasites.	Growth of pigs.
6. Day versus night collection.	Milk production of cows.
7. Temperature.	Swiftness of ants.
8. Temperature.	Egg production of chickens.
9. Temperature and source of nectar.	Rate of crystallization of honey.
10. Color of walls of cage.	Guinea pig activity levels.
11. Location (right or left) and color (black or white).	Memory in dogs (as measured by a delayed matching-to-sample procedure).
12. Various habitats of different species of worms.	Food preferences.
13. Different scents.	Prey detection by African clawed frogs.
14. Caffeine.	*Daphnia* activity levels.
15. Temperature and sky conditions (amount of sun).	Length of time for adult dragonfly to emerge from last nymphal exoskeleton.
16. Gender.	Food preference (and amounts eaten) of adult praying mantis.
17. Various samples of rotten logs differing in tree species and size.	Amount and variety of invertebrates living in them.

Appendix B

List of Suppliers

Acorn Naturalists
17300 E. 17th St., Ste. J-236, Tustin, CA 92780
(714) 838-4888, (800) 422-8886
Fax (714) 838-5869, (800) 452-2802
E-mail: Acorn@aol.com
http://www.acorn-group.com
(Field equipment for biology and earth science)

Adventure Co.
435 Main St., Johnson City, NY 13790
(607) 729-6512, (800) 477-6512
Fax: (607) 729-4820
(Equipment and materials for all the sciences)

AIMS Education Foundation
P.O. Box 8120, Fresno, CA 93747-8120
(209) 255-4094, Fax: (209) 255-6396
http://www.AIMSedu.org/
(Math and science manipulatives, including Lego products)

American 3B Scientific
2189 Flintstone Dr., Unit O, Tucker, GA 30084
(770) 492-9111, (888) 3BMODELS
http://www.A3BS.com
(Wide product range and economical prices for anatomical models)

American Science & Surplus
3605 Howard St., Skokie, IL 60076
(847) 982-0870, Fax: (800) 934-0722
(Electromechanical devices such as motors, gears, and switches; optics; raw materials including metals; surplus goods from industrial overstocks, bankruptcies)

Arbor Scientific
P.O. Box 2750, Ann Arbor, MI 48106-2750
(313) 913-6200, (800) 367-6695
(Physics materials)

Associated Microscope
P.O. Box 1076, Elon College, NC 27244
(910) 578-3893, (800) 476-3893
(Microscopes and video systems)

Azion
205 Kelsey Ln. Ste. 1, Tampa, FL 33619
(813) 621-2230, (800) 533-6444
Fax: (813) 626-5419
(Plastic beakers, cylinders, test-tube racks, and more)

Ben Meadows
3589 Broad St., Atlanta, GA 30341
(770) 455-0907, (800) 241-6401
Fax: (800) 628-2068
http://www.benmeadows. com
(Field and lab equipment for forestry, geology, horticulture, wildlife, botany, entomology, limnology, physiology)

Biology Store, The
275 Pauma Pl., P.O. Box 2691, Escondido, CA 92033
(619) 745-1445, (800) 654-0792 (CA, AZ, NV, OR, WA)
Fax: (619) 489-2268
(Living and preserved biological specimens and entomology supplies)

Bio-Rad Laboratories Life Science Group
2000 Alfred Nobel Dr., Hercules, CA 94547
(510) 741-1000, (800) 876-3425
E-mail: ron_mardigian
@bio-rad.com
(Molecular biology, electrophoresis, and chromatography)

Boreal Laboratories, Ltd.
399 Vansickle Rd., St. Catharines, ON, CANADA L2S 3T4
(905) 984-3000, (800) 387-9393
(Complete line of science supplies and equipment; over 2000 manufacturers)

Brinkmann Instruments

1 Cantigue Rd., P.O. Box 1019, Westbury, NY 11590-0207
(516) 334-7500, (800) 645-3050
Fax: (516) 334-7506
http://www.brinkmann.com
(Scientific instruments, apparatus, and disposables)

Brunton Co.

620 E. Monroe, Riverton, WY 82501
(307) 856-6559, (800) 443-4871
Fax (307) 856-1840
(Labware and apparatus)

Bureau of Economic Geology

The University of Texas at Austin, P.O. Box X, University Station,
Austin, TX 78713
(512) 471-1534
(Rocks and mineral specimens, globes, and maps)

Carolina Biological Supply Co.

2700 York Rd., Burlington, NC 27216
(910) 584-0381, (800) 334-5551
E-mail: carolina@carolina.com
http://www.carolina.com
(Thousands of products in all the sciences)

Casio, Inc.

570 North Pleasant Ave., Dover, NJ 07801
(201) 361-5400, (800) 582-2763
http://www.casio-usa.com
(Calculators and stand-alones for data collection)

CCS Educational

24 Rogate Pl., Scarborough, ON CANADA M1M 3C3
(416) 267-8844
E-mail: ccs@io.org
*(Canadian representative Vernier Software and Texas Instruments for Computer
Assisted Labs)*

Central Scientific Co.
3300 CENCO Pkwy., Franklin Park, IL 60131
(708) 451-0150, (800) 262-3626
Fax: (708) 451-0231
http://www.cenconet.com
(Full-line science supplier with physics expertise)

CHEMetrics, Inc.
Route 28, Calverton, VA 20138
(540) 788-9026, (800) 356-3072
Fax: (703) 788-4856
E-mail: mkt@chemetrics.com
http://www.chemetrics.com
(Water analysis)

Connecticut Valley Biological Supply Co.
82 Valley Rd. P.O. Box 326, Southampton, MA 01073
(413) 527-4030, (800) 628-7748
Fax: (800) 355-6813
(Biology and earth science)

Corning
Science Products Division, BF-02, Corning, NY 14831
(607) 974-0353, (800) 222-7740
(Pyrex and Corning brand laboratory glassware)

Cosmos Ltd.
9215 Waukegan Rd., Morton Grove, IL 60053
(800) 643-2351
(Microscopes, magnifiers, and telescopes)

Cynmar Corp. E & L Microscope & Balance Co.
131 N. Broad St., P.O. Box 530, Carlinville, IL 62626
(800) 223-3517
Fax: (217) 854-5154
(Videomicroscope equipment, labware, and general science supplies)

Daedalon Corp.
P.O. Box 2028, Salem, MA 01970-6228
(508) 744-5310, (800) 233-2490
Fax: (508) 745-3065
(Physics apparatus, physics-oriented software, and computer interfacing equipment)

Data Harvest Educational
349 Lang Blvd., Grand Island, NY 14072
(905) 828-6166, (800) 436-3062
Fax: (905) 607-3469
E-mail: Easylog@interlog.com
(Probeware for collecting and analyzing scientific data)

Davis Instruments
3465 Diablo Ave., Hayward, CA 94545
(510) 732-9229, (800) 678-3669
Fax: (510) 670-0589
E-mail: info@davisnet.com
http://www.davisnet.com
(Weather stations: models range from simple temperature and barometric pressure measurements to detailed wind speed and direction, wind chill, solar and UV radiation, heat index, soil temperature, leaf wetness, and rainfall measurements)

Delta Education
12 Simon St., P.O. Box 300, Nashua, NH 03061
(603) 598-7170, (800) 260-9577
(Complete line of manipulative materials for science activities)

DuPont NEN Products
549 Albany St., Boston, MA 02118
(800) 551-2121
(Biology and chemistry labware, stains, indicators, chemicals, and test kits)

E & L Instruments
70 Fulton Terr., P.O. Box 1942, New Haven, CT 06509
(203) 624-3103, (800) 368-0880
(Fiber optics; lasers; linear, digital, and AC/DC electronics)

Eagle Instruments
20 Adrian Ct., Burlington, CA 94010
(415) 697-2955, (800) 332-8288
Fax: (415) 697-9207
E-mail: eagledp@aol.com
(Labware, videomicroscopes)

Edmund Scientific Co.
101 E. Gloucester Pike, Barrington, NJ 08007
(609) 547-8880
E-mail: scientifics@edsci.com
(Discrepant event items, microscopes, lasers, balances, and videomicroscope systems)

Educational Rocks and Minerals
P.O. Box 60574, Florence, MA 01060
(410) 586-6068
(Rocks and minerals)

Edvotek
P.O. Box 1232, West Bethesda, MD 20827-1232
(301) 251-5990, (800) 338-6835
E-mail: edvotek@aol.com
(Reagents, experiment kits, and equipment for biotechnology)

EmCal Scientific
P.O. Box 27105, San Diego, CA 92198-1105
(619) 486-0207, (800) 872-2397
(Microscope accessories)

Entomological Society of America
9301 Annapolis Rd., Lanham, MD 20706
(301) 731-4535
(Prepared slides of insects)

Eric Armin Inc. (EAI)
P.O. Box 644, Franklin Lakes, NJ 07417
(201) 891-9466, (800) 272-0272
Fax: (201) 891-5689
(Calculators, stopwatches and other timing devices)

ERIM-Earth Observation Group
P.O. Box 134001, Ann Arbor, MI 48113-4001
(313) 994-1200 ext. 3350
(Earth-science equipment and supplies)

Eshed Science & Technology
445 Wall St., Princeton, NJ 08540
(609) 683-4884, (800) 777-6268
(General science supplies)

Estes Industry
1295 H St., Penrose, CO 81240
(719) 372-6565, (800) 820-0202
Fax: (800) 820-0203
(Complete line of model rocket kits, engines, launch equipment, and accessories)

ETA
620 Lakeview Pkwy., Vernon Hills, IL 60061-9923
(847) 816-5050, (800) 445-5985
Fax: (800) ETA-9326 (orders)
E-mail: info@etauniverse.com
http://www.eta.universe.com
(Extensive range of science materials)

Etgen's Science Stuff
3600 Whitney Ave., Sacramento, CA 95821
(916) 972-1871
(Authentic museum-quality castings of famous fossils)

Export Science Associates
36 Lexington Ave., Suffern, NY 10901
(914) 357-0598
E-mail: roger@icu.com
(Small-scale chemistry supplies, tie-dye materials, science toys, and other chemistry-related materials)

Extech Instruments Corp.
335 Bear Hill Rd., Waltham, MA 02154
(617) 890-7440
(Portable and bench-top meters and sensors for measuring pH, conductivity, temperature, RPM, and weight)

Fisher Scientific
485 S. Frontage Rd., Educational Materials Division, Burr Ridge, IL 60521
(708) 655-4410, (800) 955-1177
(Supplies for all the sciences including new technology products, formaldehyde-free specimens, and electrophoresis apparatus)

Flight Systems, Inc.
9300 E. 68th, Raytown, MO 64133
(816) 697-2011
(Rockets, motors, complete kits, launch stands)

Flinn Scientific
Raddant Road, P.O. Box 219, Batavia, IL 60510
(630) 879-6900, (800) 452-1261
Fax: (630) 879-6962
E-mail: flinnsci@aol.com
(Specializes in supplies for biology and chemistry, laboratory chemistry safety, computer interface systems, and probeware)

Flowerfield Enterprises
10332 Shaver Rd., Dept. 105, Kalamazoo, MI 49002
(616) 327-0108
Fax: (616) 327-7009
(Worm-away vermicomposting kits and live worms)

Forestry Suppliers
P.O. Box 8397, Jackson, MS 39284-8397
(601) 354-3565, (800) 647-5368
Fax: (800)543-4203
E-mail: fsicats@teclink.net
(Orienteering compasses, testing instruments for water and soil sampling, tree borers, diameter tapes, field scales, entomology collecting supplies, rock picks, and sieves)

FOTODYNE
950 Walnut Ridge Dr., Hartland, WI 53029-9388
(414) 369-7000, (800) 362-4657
E-mail: FOTODYNE@aol.com
(Electrophoresis and photodocumentation equipment including gel boxes, transilluminators, thermal cyclers, and camera and power supplies)

Frey Scientific, Div. of Beckley Cardy
100 Paragon Pkwy., Mansfield, OH 44903
(419) 589-1900, (800) 225-FREY ext. 3739
http://www.BeckleyCardy.com
(Supplies and equipment in all sciences but especially chemistry)

General Supply Corp.
P.O. Box 9347, Jackson, MS 39286-9347
(601) 981-3882, (800) 553-2457
(Field studies equipment, including portable test kits for sampling air, soil, and water)

Hach Co.
P.O. Box 389, Loveland, CO 80539
(303) 669-3050, (800) 227-4224
(Dozens of test kits, portable and laboratory instruments, reagents, and labware)

Halsteducate Co.
2140 Lincolnwood Dr., Evanston, IL 60201-2061
(847) 866-6744
(Stretch and take-apart models: atoms, molecules, ionic compounds, atomic orbitals)

Hampden Engineering Corp.

99 Shaker Rd., P.O. Box 563, E. Longmeadow, MA 01028-0563
(413) 525-3981, (800) 253-2133
(Labware for chemistry and physics, including meters, gauges, and rulers)

Hands-On Science

6541 E. 40th St., Unit M, Tulsa, OK 74145-4521
(918) 664-4561
(Supplies for chemistry, physics, and earth science)

Hanna Instruments

584 Park East Dr., Woonsocket, RI 02895
(800) 426-6287
(ISO 9001–registered producer of water-quality measurement and control instrumentation for lab and field)

Ideal School Supply Co.

5623 W. 115th St., Alsip, IL 60482
(708) 385-0400, (800) 323-5131
(Limited supplies in biology, chemistry, physics, and earth science)

Innova Corp.

115 George Lamb Rd., Leyden, MA 01337
(413) 624-0102, (800) 546-6682
(20-inch and 30-inch terrestrial globes that show land and seabed features in detailed and accurate relief)

Intelitool

P.O. Box 459, Batavia, IL 60510
(630) 406-1041, (800) 227-3805
Fax: (630) 405-1079
E-mail: info@intelitool.com
http://www.com/intelitool/
(Computer-interfaced data acquisition/analysis hardware/software packages for physiology: electrocardiography, reflex arc, spirometry, muscle physiology, and exercise science)

International Optics PLC

325 Northgate Cr., P.O. Box 2050, Warrendale, PA 15086
(412) 934-2810
Fax: (412) 934-2811
(Microscopes and video equipment to be used with or without microscopes)

JT&A
4 Herbert St., Alexandria, VA 22305
(703) 519-2180
Fax: (703) 519-2190
http://www.jtainc.com
(Tabletop models of watersheds)

Kimble/Kontes
1025 Spruce St., Vineland, NJ 08360
(609) 692-8500, (800) 223-7150
Fax: (609) 692-3242
E-mail: kimkon@acy.cligex.net
http://www.kimble-kontes.com
(Macro- and microscale chemistry kits and accessories, reusable and disposable chemistry glassware)

Kons Scientific Co., Inc.
P.O. Box 3, Germantown, WI 53022-0003
(414) 242-3636
Fax: (414) 677-3435
E-mail: vmgw62@prodigy.com
(Models, living materials, apparatus, safety, health, and physical science equipment)

Laboratory Craftsmen, Inc.
P.O. Box 148, Beloit, WI 53512
(608) 362-2255, (800) 476-8802
Fax: (608) 362-2255
(Mini hot plates and ceramic heating mantles, magnetic stirrers)

Lab Safety Supply
401 W. Wright Rd., P.O. Box 1368, Janesville, WI 53547-1368
(608) 754-2345, (800) 356-0783
(Lab safety supplies and equipment)

Lafayette Instrument
3700 Sagamore Pkwy., P.O. Box 5729, Lafayette, IN 47903
(317) 423-1505, (800) 428-7545
(Physiological recording, biofeedback, reaction timing)

LaMotte Co.
P.O. Box 329, Chestertown, MD 21620
(410) 778-3100, (800) 344-3100
Fax: (410) 778-6394
E-mail: ese@lamotte.com
http://www.lamotte.com/ese
(Test kits, sampling equipment, and electronic instruments for analysis of water, soil, and air)

Learning Alternatives
2370 W. 89A-Ste. #5, Sedona, AZ 86336
(602) 204-2172, (800) HANDS-ON
(Equipment and supplies in all sciences)

Learning Things
68A Broadway, P.O. Box 436, Arlington, MA 02174
(617) 646-0093
Fax: (617) 646-0135
(Designs, manufactures, and distributes own equipment and supplies in all sciences)

Leica
111 Deerlake Rd., Deerfield, IL 60015
(847) 405-0123, (800) 248-0123
http://www.leica.com
(Microscopes and microscope imaging)

Let's Get Growing!
1900 Commercial Wy., Santa Cruz, CA 95065
(800) 408-1868
Fax: (408) 464-1868
E-mail: letsgetgro@aol.com
http://www.letsgetgrowing.com
(Worms, bats, butterflies, minerals, bugs, weather, and more)

Life Technologies
P.O. Box 6009, Gaithersburg, MD 20884-9980
(800) 828-6686
(Over 3000 products for molecular biology, cell culture, and cell biology)

Lyon Electric Co.

2765 Main St., Chula Vista, CA 91911
(619) 585-9900
E-mail: lyonelect@aol.com
*(See-through bird and reptile incubators,
animal intensive care units)*

Magnet Source Master Magnetics, The

607 S. Gilbert, Castle Rock, CO 80104
(303) 688-5303, (800) 525-3536
(Horseshoes, wands, marbles, magnetizers, and more)

Medical Plastics Laboratory

P.O. Box 38, Gatesville, TX 76528
(817) 865-7221, (800) 433-5539
(Anatomical reproductions and skeletons of bone and plastic)

Meiji Techno America

2186 Bering Dr., San Jose, CA 95131
(401) 428-0472, (800) 832-0060
(Microscopes, videomicroscopes, and accessories)

Metrologic Instruments

90 Coles Rd., Blackwood, NJ 08012
(602) 228-8100
Fax: (609) 228-6673
*(Eleven models of low-power lasers, laser power meters, and photometers; laser
kits and accessories)*

Micro-Mole Scientific

1312 N. 15th, Pasco, WA 99301
(509) 545-4904
E-mail: mauch1312@aol.com
(Microscale labware and plasticware)

Micro-Optics

68-21 Fresh Meadow Ln., Fresh Meadows, NY 11365
(718) 961-8833, (800) 776-1771
(Microscopes, balances, videomicroscopes, and digital imaging)

Mountain Home Biological
P.O. Box 1142, White Salmon, WA 98672
(509) 493-2669
Fax: (509) 493-4321
(Owl pellets, skull sets, Riker mounts)

Museum Products Co.
84 Route 27, Mystic, CT 06355
(860) 536-6433, (800) 395-5400
Fax: (860) 572-9589
(Binoculars, compasses, Toys in Space, shark teeth, and seashells)

NADA/Nakamura Scientific Ltd.
39 Butternut St., P.O. Box 1336, Champlain, NY 12919
(518) 298-2393, (800) 799-NADA
E-mail: nadasci@slic.com
http://www.nadasci.com
*(Over 10,000 science equipment and supply products
including hand-held generators)*

Naige Nunc International
75 Panorama Crk. Dr., P.O. Box 20365
Rochester, NY 14602
(716) 586-3294, (800) 625-4327
*(Plastic burettes, droppers, bottles, hand vacuum pumps
and chambers, safety shields)*

Nasco
901 Janesville Ave., P.O. Box 901, Fort Atkinson, WI 53538-0901
(414) 563-2446, (800) 550-9595
Fax: (414) 563-8296
E-mail: info@nascofa.com
(Comprehensive offering of supplies and equipment in all sciences)

NASCO-Modesto
4825 Stoddard Rd., P.O. Box 3837, Modesto, CA 95352-3837
(209) 545-1600, (800) 558-9595
E-mail: modesto@nascofa.com
http://www.nascofa.com
*(Specializes in live and preserved specimens as well as
offering products for all sciences)*

National Association of Conservation Dist

408 E. Main, P.O. Box 855, League City, TX 77574-0855

(713) 332-3402, (800) 825-5547

(Acreage measuring device; residue measurement guide; conservation plants, trees, and shrubs)

National Gardening Association

180 Flynn Ave., Burlington, VT 05401

(802) 863-1308, (800) 863-5962

E-mail: nga@garden.org

http://www.garden.org

(Indoor gardening equipment, including light gardens, hydroponics units, worm bins, and greenhouses)

Nature Discoveries

389 Rock Beach Rd., Rochester, NY 14617

(716) 544-8198

(Butterfly and moth stages, butterfly rearing equipment)

Nature Store

455 Cedar St., West Barnstable, MA 02668

(508) 362-6429

(Owl pellets)

Nebraska Scientific

3823 Leavenworth St., Omaha, NE 68105-1180

(402) 346-7214, (800) 228-7117

Fax: (402) 346-2216

(Preserved specimens: fetal pigs, frogs, brains, eyes, hearts)

NEN Life Sciences

549 Albany St., Boston, MA 02118

(617) 542-9595, (800) 551-2121

Fax: (800) 772-2NEN

(Specialty reagents, supplies and kits for biotechnology)

New Omni Inc.

P.O. Box 450448, Atlanta, GA 30345

(800) 472-0150

(Microscopes and magnifiers)

Newport Corp.
1791 Deere Ave., Irvine, CA 92714
(714) 863-3144, (800) 222-6440
(Optics, fiber optics, photonics)

Nikon, Inc. Instrument Group
1300 Walt Whitman Rd., Melville, NY 11747-3064
(516) 547-8500
(Microscopes and videomicroscopes)

Northwest Laboratory Supply
5510 Nielson Rd., #B, Ferndale, WA 98248
(360) 384-1673
(Microscopes, balances, pH meters, Lego dacta, labware, and models)

Numberg Scientific Co.
6310 SW Virginia Ave., Portland, OR 97201
(503) 246-8297, (800) 527-8594 (in OR), (800) 826-3470 (outside OR)
Fax: (503) 246-0360
(Microscopes and supplies for all sciences)

Ocean Biologics, Inc.
5315 139th Pl., SW, Edmonds, WA 98026
(206) 743-0894, (800) 647-9721
Fax: (206) 743-0894
(Supplies/equipment for electrophoresis and other biotechnology)

Ohaus Corp.
29 Hanover Rd., Florham Park, NJ 07932
(201) 377-9000, (800) 672-7722
Fax: (201) 593-0359
http://www.ohaus.com
(Individual mass sets and Ohaus electronic analytical, top-loading, portable, and mechanical balances)

Omnion
P.O. Box O, Rockland, MA 02370
(617) 878-7200
Fax: (617) 878-7465
(Chemistry labware)

Omni Resources
1004 S. Mebane St., P.O. Box 2096, Burlington, NC 27216-2096
(800) 742-2677
Fax: (910) 227-3748
(Earth science supplies)

Other Worlds Enterprises
2029 Sunshine Circle, P.O. Box 6193, Woodland Park, CO 80866-6193
(719) 687-3840
(The Great Rock Swap, a unique exchange opportunity to obtain quality rock specimens from around the country at low cost)

Owls, Etc.
15 Riviera Ct., Great River, NY 11739
(516) 581-5336
Fax: (516) 581-1905
(Owl pellets)

Oxford Instruments Nuclear Measurements Group
601 Oak Ridge Tpke., P.O. Box 2560, Oak Ridge, TN 37831-2560
(615) 483-8405, (800) 769-3673
(Educational nuclear science instrumentation for measuring radioactive material and radioisotopes; also manufactures radioisotopes)

Parco Scientific Co. Instrument Group
316 Youngstown-Kingsville Rd., SE, P.O. Box 189, Vienna, OH 44473
(216) 394-1100, (800) 24-PARCO
(Microscopes)

PASCO Scientific
10101 Foothills Blvd., Box 619011, Roseville, CA 95678-9011
(916) 786-3800, (800) 772-8700
(Physics supplies including interfacing hardware and accessories)

Patio Garden Pond
7919 S. Shields, Oklahoma City, OK 73149
(405) 634-POND, (800) 487-7663
(Aquaria and water garden products)

Pellets
P.O. Box 5484, Bellingham, WA 98227-5484
(360) 733-3012
Fax: (360) 738-3402
(Barn owl pellets)

Protein Solutions, Inc.
6009 Highland Dr., Salt Lake City, UT 84121
(801) 583-9301
(Specializes in bioluminescence)

Quest Aerospace Education
519 W. Lone Cactus Dr., Phoenix, AZ 85027-2921
(602) 582-3438, (800) 858-7302
Fax: (602) 582-3828
(Model rocket kits, rocket motors, launch control systems)

RCR Scientific
206 W. Lincoln Ave., Goshen, IN 46526
(219) 533-3351
(Microbiology)

Resolution Technology
26000 Avenida, Aeropuerto #22, San Juan Capistrano, CA 92675
(714) 661-6162
Fax: (714) 661-0114
(Videomicroscope systems)

Rocket Age Enterprises
9 Lance Rd., Lebanon, NJ 08833
(908) 439-3559
Fax: (908) 439-2140
(Model rockets, engines, launch systems, mouse trap racer kits, and more)

Sargent-Welch/VWR Scientific
911 Commerce Ct., Buffalo Grove, IL, 60089-2375
(800) SAR-GENT
E-mail: sarwel@sargentwelch.com
http://www.SargentWelch.com
(Comprehensive offerings of products in all sciences. On-line Internet catalog solely devoted to science-education products)

Satellite Data System
P.O. Box 219, Cleveland, MN 56017-0219
(507) 931-4849
E-mail: sds@ic.mankato.mn.us
http://ic.mankato.mn.us/-sds
(Satellite demodulator boards for receiving weather satellite images on IBM-PC compatibles, plus accessories)

Schoolmasters Science
745 State Circle, P.O. Box 1941, Ann Arbor, MI 48106
(313) 761-5072 (800) 521-2832
Fax: (800) 654-4321
E-mail: schscience@aol.com
http://www.schoolmasters.com
(Products for all sciences)

Science First
95 Botsford Pl., Buffalo, NY 14216
(716) 874-0133, (800) 875-3214
Fax: (716) 874-9853
(Physics, math, technology products, including electronic timers and photogates)

Science Import
1990 Boul. Charest Quest, Ste. 106, Quebec City, Quebec, CANADA
G1N 4K8
(418) 527-1414
Fax: (418) 527-1970
(Periodic charts, lasers, forces tables, eye models, photogate timers)

Science Instruments Co.
6122 Reisterstown Rd., Baltimore, MD 21215
(410) 358-7810
(Battery-operated equipment integrates computer techniques with instrumentation)

Science Kit and Boreal Laboratories
777 E. Park Dr., Tonawanda, NY 14150
(716) 874-6020, (800) 828-7777
Fax: (800) 828-3299
http://sciencekit.com
(Comprehensive offerings in all sciences)

Science Source, The
P.O. Box 727, Waldoboro, ME 04572
(207) 832-6344, (800) 299-LINX
E-mail: scisourc@midcoast.com
http://www.midcoast.com/-scisourc
(Physics/physical science apparatus, LINX building/construction system)

Science Supply Co.
P.O. Box 836, Yarmouth, ME 04096
(800) 277-6520
(Limited products in biology, chemistry, physics, and earth science)

SCI-MA Education
325 S. Westwood, #4, Mesa, AZ 85210
(602) 464-5605, (800) 848-6284
(Kits of hands-on products in biology, chemistry, physics, and earth science)

Sci Space Craft International
Catalina Station, P.O. Box 61027, Pasadena, CA 91116-7207
(818) 792-4300, (800) 472-4548
(Economical models of space exploration machines)

Seiler Instrument Co.
170 E. Kirkham Ave., St. Louis, MO 63119-1791
(314) 968-2282, (800) 489-2282
Fax: (314) 968-2637
(Microscopes and planetarium projectors)

Skilcraft
328 N. Westwood, Toledo, OH 43607
(419) 536-8351
(Test kits and rock/mineral samples)

Skullduggery
624 S. B St., Tustin, CA 92680
(714) 832-8488
Fax: (714) 832-1215
(Fossil replicas and anthropological models)

Skulls Unlimited International
P.O. Box 6741, Moore, OK 73153
(405) 632-4200, (800) 659-7585
(Over 200 different natural bone skeletons from around the world and 50 different replica skulls of endangered species)

Southern Precision Instruments
3419 E. Commerce St., San Antonio, TX 78220
(512) 224-5801, (800) 888-5801 ext. 202
(Microscopes, microprojectors, and prepared slide sets)

Southland Instruments, Inc.
P.O. Box 1517, Huntington Beach, CA 92647
(714) 847-5007, (800) 862-0447
(Microscopes and accessories, balances)

Spectronic Instruments
820 Lincoln Ave., Rochester, NY 14625
(716) 248-4000, (800) 654-9955
Fax: (716) 248-4014
E-mail: info@spectronic.com
http://www.spectronic.com
(Spectrophotometers)

Swift Instruments
P.O. Box 562, San Jose, CA 95106
(408) 293-2380, (800) 523-4544
(Microscopes, videomicroscopes, hand microtomes, binoculars, telescopes, weather instruments)

Synthephytes
P.O. Box 1032, Angleton, TX 77516-1032
(713) 369-2044
(Plant biotechnology)

TEDCO
498 S. Washington St., Hagerstown, IN 47346
(317) 489-4527, (800) 654-6357
Fax: (317) 489-5752
(Gyroscopes, prisms, science kits)

Thornton Educational Products
P.O. Box 2566, Naples, FL 33939
(800) 64T-EPCO
(Modular electronic equipment and computer interface hardware and software)

Tiops
P.O. Box 10852, Pensacola, FL 32524
(904) 479-4415, (800) 200-3466
Fax: (904) 479-3315
(Kits on fossils and gems, crustaceans, killifish hatching)

Tout About Toys, Inc.
551 Foster City Blvd., Suite 1, Foster City, CA 94404
(415) 286-0214, (800) 598-1523
(Math and science manipulatives)

Triarch
P.O. Box 98, Ripon, WI 54971-0098
(414) 748-5125, (800) 848-0810
Fax: (414) 748-3034
(Prepared microscope slides, microscopes and videomicroscopes)

Tri Space
P.O. Box 7166, McLean, VA 22106-7666
(703) 442-0666
Fax: (703) 442-9677
(Weather systems, satellite weather systems for IBM compatibles)

Trippensee Planetarium Co.
301 Case St., Saginaw, MI 48602-2097
(517) 799-8115, (800) 799-8115
(Multiple-motion planetaria, galaxy models)

Ulrich's Fossil Gallery
Fossil Station #308, Kemmerer, WY 83101
(307) 877-6466
Fax: (307) 877-3289
(Middle Eocene fossil fish preparation kits)

Uptown Sales
33 N. Main St., Chambersburg, PA 17201
(800) 548-9941
Fax: (717) 264-8123
http://www.hobbyplace.com
(Estes model rocketry, plastic anatomy models, insect collecting, model finishing materials)

U.S. Fish & Wildlife Service
National Wetlands Inventory, 9720 Executive Center Dr.
St. Petersburg, FL 33702
(813) 570-5412
Fax: (800) USA-MAPS
(Wetlands maps of the U.S., digital data also available for some parts of the U.S.)

Vernier Software
8565 SW Beaverton-Hillsdale Hwy., Portland, OR 97225-2429
(503) 297-5317
E-mail: cvernier@vernier.com
http://www.vernier.com
(Computer interfacing hardware and software for measuring temperature, pH, voltage, frequency, etc.)

WARD's Natural Science Est.
5100 W. Henrietta Rd., P.O. Box 92912, Rochester, NY 14692-9012
(716) 359-2502, (800) 962-2660
Fax: (800) 635-8439
http://www.wardsci.com
(Comprehensive offerings in all sciences)

WESCO
1577 Colorado Blvd., Los Angeles, CA 90041
(213) 257-0832, (800) 48-WESCO
(Microscopes)

West Coast Aquatics
906 Calle Collado, Thousand Oaks, CA 91360
(805) 499-7866
Fax: (805) 499-3637
(Aquaria and accessories, live temperate animals from the Pacific coast)

Wildlife Supply Co.
301 Cass St., Saginaw, MI 48602-2097
(517) 799-8100
(Aquatic sampling equipment)

Wind & Weather
P.O. Box 2320-ST, Mendocino, CA 95460
(707) 964-1284, (800) 922-9463
Fax: (707) 964-1278
(Barometers, thermometers, hygrometers, psychrometers, wind direction instruments, anemometers, rain gauges, and more)

Women in Mining Education Foundation
1801 Broadway, Ste. 400, Denver, CO 80202
(303) 298-1535
(Students can receive free mineral samples by written request)

Young Entomologists' Society
1915 Peggy Pl., Lansing, MI 48910-2553
(517) 887-0499
Fax: (517) 887-0499
E-mail: yesbugs@aol.com
http://insect.ummz.lsaumich.edu/yes/yes.html
(Materials on insects, spiders, snails, and worms)

Young Naturalist Co.
1900 N. Main, Newton, KS 67114
(316) 283-4103
Fax: (316) 283-9108
(Leaf, seed, crop, and twig identification kits)

Appendix C

Tips for Teachers

In the past, science fairs have been a mainstay of many science programs. Today, however, their presence and implementation are even more important than ever. A careful reading of the National Science Education Standards[1] makes explicit the valuable experience science fairs provide. In its explanation of the standard for science as inquiry, the National Standards state: "Students . . . should have the opportunity to use scientific inquiry and develop the ability to think and act in ways associated with inquiry, including asking questions, planning and conducting investigations, using appropriate tools and techniques to gather data, thinking critically and logically about relationships between evidence and explanations, constructing and analyzing alternative explanations, and communicating scientific arguments."[2] These very words could be used to describe what a science fair project is all about.

---

[1] National Research Council. *National Science Education Standards.* Washington, D.C.: National Academy Press, 1996.

[2] Ibid.

Our middle school students are capable of real, investigative science projects. For middle school students, real science is the performance of science experiments with data collection: They have a natural enthusiasm for this work. To assign a science project that is merely the making of a model or the writing of a report is less than second best: It is a turnoff. Also, science projects that lack an experiment and data collection are rarely, if ever, given recognition at district and state science competitions. Therefore, to assign a student such a project is the same as setting that student up for a disappointment.

On the other hand, to supervise science projects involving data collection as well as the other necessary parts is a lot of extra work for the teacher. Those teachers willing to do it (and I hope that you are one of them) should be commended for their willingness to put forth the extra effort to make science real and exciting for their students. I hope that this book has already made or soon will make that task easier for you.

If besides teaching science and overseeing the science fair projects you also are the director of your local science fair, I doubly commend you. I also will share my secret for coping with the almost overwhelming number of details that being fair director/teacher/supervisor of projects brings: *Spread out the necessary tasks for pulling off a successful science fair over a whole school year's time.*

Spring

Start preparing for next year's science fair in the spring. You will need the biggest space available for your fair so that your students will not be elbow to elbow. Usually, the biggest space in a school is its gym. Schedule your use of it now, because this is when the athletic director is setting up the basketball schedule for next year. Talk to the athletic director and arrange a night for the gym to be free and for your students not to be involved with any athletic events. Next, explain to the gym teacher that you will need to set up your projects on that day and offer to trade places with him or her. Finally, make a building request in writing to your principal for both the gym and the cafeteria on the evening of your science fair. (You can use the cafeteria as a hospitality center for judges and parents.) Include in your building request a map of how the gym should be set up with enough table space for all the projects and a microphone so that you can make any needed announcements.

Your task in guiding your students through science fair projects is made harder or easier, depending on the preparation they have had before coming to you. Now is the time to urge the teachers who have your future students to assign a research report complete with a bibliography. (It is helpful when your incoming students know the difference between a biography and a bibliography!) Then you can build on the skills your students bring with them instead of having to start from scratch. Trying simultaneously to teach both the skill of scientific methods and concepts *plus* the skill of research report writing can be overwhelming for both the teacher and the students.

Having to teach all the skills necessary to put together a science fair project brings up another point: You could share the fun, glory, and work of the fair with other teachers, including nonscience teachers. Instead of just you overseeing all the projects and teaching all the skills, you could choose one of these options: (1) Have an interdisciplinary science fair in which all students would do a science project but the library research would be a social studies assignment, the backdrop an art assignment, the graphs a math assignment, and so on, or (2) have an academic fair for your grade level in which all the teachers cooperate and students may choose to do either a math, social studies, science, or language arts project.

Spring is also a good time to stir up enthusiasm and anticipation among your prospective students. Visit the classrooms of your prospective students and tell them about the science fair. I bring along some of my most successful seventh graders to demonstrate their projects to the sixth grade. I also share our "brag book" of pictures, newspaper clippings, and other evidence of the recognition my students have received at district and state fairs.

Autumn

Autumn is the head start you need to lay the groundwork to make the fair go smoothly. Stress hands-on lab experiences with data collection in your science classes. This both reinforces concepts and helps students learn about scientific processes in a concrete manner.

Students will know how to write a summary for their experiment by the time they begin working on their projects if you require them to write up their autumn lab experiments in science fair form. Make sure they have all the parts: topic, question, hypothesis, materials, procedures, results in chart or graph form, and conclusion.

If you are the person responsible for helping students on their research papers, reserve public library and school library time for students to be shown where the reference materials can be found. Also, arrange to have reference materials in the classroom available for your students.

The Internet is a primary resource for up-to-date information. By helping students become familiar with using the Internet in the autumn, they will know how to wield this powerful tool when it comes time to work on their science fair project in the winter. The Internet has direct links to libraries, museums, universities, government agencies, and world organizations. Make access to this wealth of information easier for your students by setting up home pages for different science topics to guide their inquiry. Start by identifying the general themes needed to develop a knowledge base that would support your science course objectives. Next, identify the topics needed for student success within those themes. Finally, determine which Internet sites are reliable (Appendix E in this book should help) and evaluate those sites for their usefulness. Develop an address list of the top sites and transfer them (i.e., set bookmarks) onto the home pages.

Communication between home and school is vital when working with middle school students. Start early to make the necessary contacts. For example, prepare a letter for parents stating the six requirements that students must meet as a part of their science fair project: the performance of an experiment with data collection, a format summary of the experiment, a research report with bibliography, a visual backdrop, an oral presentation, and attendance on the night of the fair. Also, write out a week-by-week schedule for students telling them what they should be working on and what due dates they should meet. (The schedule provided at the end of "Step 1: Getting Organized for the Tasks Ahead" could help you do this.) Have this letter and schedule ready for parent conferences in the autumn so that you have concrete information to hand your parents.

Now is also the time to prepare judging sheets. In order to be as fair as possible to my students, I base my judging sheets on the six things that I have required them to do for their science fair project (listed earlier). I assign the most possible points to a well-designed and well-executed experiment because I feel that the experiment is the most important part of their project.

Other details that you could get out of the way in the fall include preparing award certificates (do you have any student calligraphers?), ordering ribbons, and reserving space and time for the awards assembly.

Winter

During student preparation of the science fair projects, give it your all. Since you have already attended to many of the details of organizing a fair, you are free to concentrate on following the week-by-week schedule. This includes anticipating students' needs to learn new skills. For example, teach them how to write a bibliography about a week before the due date for bibliographies. If your students need information that is outside your expertise, help them contact resource people in the community who can talk to them.

Several weeks before the fair, line up your judges. Personal contact by telephone works best. I put a high premium on student-scientist interaction, so each student is judged by at least two judges. On the other hand, I treat my judges well and do not assign them to more than four or six projects each.

In the ten days leading up to the fair, do not panic! People will often do wonders when the time crunch is on!

Do go over the judging sheets in class. This helps students know what to expect. Also, they can save you valuable time by filling out their name, the date, and the title and number of their project at the top of the judging sheets. Next, make up the judging assignments and group the sheets together for each judge so that you are ready for the day of the fair. Finally, make up name tags for the judges, arrange to have a volunteer photographer, and alert the media (newspapers, radio, even local TV).

On the Actual Day of the Fair and Afterward

After your students have set up their projects during their class periods, you could invite other grades to view the projects with your students serving as hosts. Be sure to remind your students to dress well for the fair and to be polite.

During the fair in the evening, be sure you have delegated as much responsibility as possible. This involves more people in the fair and leaves you free to troubleshoot. It also leaves you free to greet your judges and enjoy yourself. (Yes, believe it or not, when the fair finally arrives, you can enjoy yourself. Students will be dressed up and on their best behavior. The whole gymnasium will have a holiday air like an old-fashioned country fair.)

After your local fair, you will have a few things to do to finish it up before you begin preparation for the district and, hopefully, state fairs. You will need to average the judges' scores, fill in names on the award certificates, and host the awards assembly. You also will need to write articles for the local newspapers and have your students write thank-you notes to the judges.

District and state fair preparation begins with drawing together a team of student volunteers who did well in the local science fair. Structure and communication between home and school are still essential for each of these team members. During the conference, read the judges' comments and draw up a contract for the tasks that the student agrees to do as part of the school's science team. These tasks should include preparing the oral presentation for videotaping and practicing it in front of the lower grades when you make your spring visits. This completes the yearly cycle that began the previous spring.

Appendix D

Further Reading

Bingham, Jane. *Science Experiments.* London, Eng.: Usborne Publishing Ltd, 1991.

Bonnet, Bob, and Dan Klen. *Science Fair Projects: The Environment.* New York: Sterling Publishing Co., Inc., 1995.

———. *Science Fair Projects with Electricity and Electronics.* New York: Sterling Publishing Co., Inc., 1996.

Brisk, Marion A., Ph.D. *1001 Ideas for Science Projects.* 2nd ed. New York: Macmillan, Inc., 1994.

Cortright, Sandy, and Will Pokriots. *Making Backyard Birdhouses: Attracting, Feeding and Housing Your Favorite Birds.* New York: Sterling Publishing Co., Inc., 1996.

Dorris, Ellen. *Entomology.* New York: Thames and Hudson, 1993.

———. *Invertebrate Zoology.* New York: Thames and Hudson, 1993.

Duensing, Edward. *Talking to Fireflies, Shrinking the Moon: Nature Activities for All Ages.* Golden, Colo.: Fulcrum Publishing, 1997.

Dyer, Alan, and Terence Dickenson. *The Backyard Astronomer's Guide.* Willowdale, Ont., Can.: Firefly Books Ltd., 1991.

Earthworks Group, The. *The Next Step: 50 More Things You Can Do to Save the Earth.* Kansas City, Mo.: Andrews and McMeel, a Universal Press Syndicate Co., 1991.

Gardner, Robert. *Science Fair Projects—Planning, Presenting, Succeeding.* Springfield, N.J.: Enslow Publishers, Inc., 1999.

Hamburg, Michael. *Astronomy Made Simple.* New York: Doubleday, 1993.

Harrington, Philip S. *Touring the Universe Through Binoculars: A Complete Astronomer's Guide.* New York: John Wiley and Sons, Inc., 1990.

Hixson, B. K. *Bernoulli's Book.* Salt Lake City, Utah: The Wild Goose Co., 1991.

Holden, Alan, and Phylis Morrison. *Crystals and Crystal Growing.* Cambridge, Mass.: MIT Press, 1982.

Iritz, Maxine Haren. *Blue-Ribbon Science Fair Projects.* New York: McGraw-Hill, 1991.

Johnson, Cathy. *The Nocturnal Naturalist: Exploring the Outdoors at Night.* Old Saybrook, Conn.: The Globe Pequot Press, 1989.

Lawlor, Elizabeth P. *Discover Nature at Sundown.* Mechanicsburg, Pa.: Stackpole Books, 1995.

MacRobert, Alan M. *Star-Hopping for Backyard Astronomers.* Cambridge, Mass.: Sky Publishing Corp., 1993.

Matloff, Gregory L. *Telescope Power: Fantastic Activities and Easy Projects for Young Astronomers.* New York: John Wiley and Sons, Inc., 1993.

Moje, Steven W. *Paper Clip Science: Simple and Fun Experiments.* New York: Sterling Publishing Co., Inc., 1996.

Myers, Norman, ed. *Gaia: An Atlas of Planet Management.* New York: Doubleday, 1993.

Rezendes, Paul. *Tracking and the Art of Seeing.* Charlotte, Vt.: Camden House Publishing, Inc., 1992.

Roberts, Royston M., and Jeanie Roberts. *Lucky Science: Accidental Discoveries from Gravity to Velcro.* New York: John Wiley and Sons, 1995.

Rosner, Marc Alan. *Science Fair Success Using the Internet.* Springfield, N.J.: Enslow Publishers, Inc., 1999.

Rubin, Louis D., Sr., and Jim Duncan. *The Weather Wizard's Cloud Book.* Chapel Hill, N.C.: Algonquin Books, 1989.

Scriven, Dorene H. *Bluebird Trails: A Guide to Success.* Minneapolis, Minn.: Bluebird Recovery Committee of the Audobon Chapter of Minneapolis, 1993.

Seed, Deborah. *Water Science.* Reading, Mass.: Addison-Wesley Publishing Co., 1992.

Self, Charles R. *Super Simple Birdhouses You Can Make.* New York: Sterling Publishing Co., Inc., 1995.

Smolinski, Jill. 50 *Nifty Super Science Fair Projects.* Chicago, Ill.: RGA Publishing Group, Inc., 1995.

Stermer, Dugald. Vanishing Flora, *Endangered Plants Around the World.* Abrams, 1995.

Sutton, Patricia, and Clay Sutton. *How to Spot an Owl.* Shelburne, Vt.: Chapters Publishing Ltd., 1994.

Tomecek, Steve. *Bouncing and Bending Light.* New York: Walt Freeman and Co., 1995.

VanCleave, Janice. *A+ Projects in Chemistry: Winning Experiments for Science Fairs and Extra Credits.* New York: John Wiley and Sons, 1993.

———. *Guide to the Best Science Fair Projects.* New York: John Wiley and Sons, Inc., 1997.

Vecchione, Glen. *100 Amazing Make-It-Yourself Science Fair Projects.* New York: Sterling Publishing Co., Inc., 1995.

Walpole, Brenda. *175 Science Experiments To Amuse and Amaze Your Friends.* New York: Random House, 1988.

Williams, Jack. *USA Today: The Weather Book.* New York: Random House, 1992.

Wingate, Philippa. *Essential Physics.* London, Eng.: Usborne Publishing Ltd, 1991.

Appendix E

Internet Addresses

Astronomy and Engineering

Carnegie Science Center
http://www.csc.clpgh.org/
> *For a skywatch hotline, an astronomical calendar, and further information on the planetarium at the Carnegie Science Center in Pittsburgh, Pennsylvania, try this site.*

Mount Wilson and Telescopes in Education (TIE)
http://tie.jpl.nasa.gov/tie/index.html
> *Learn all about the telescope, or you can take a complete tour of Mount Wilson—America's Observatory. The TIE program offers free observing time to schools on weekdays and time for a fee to amateurs on weekends and holidays. The telescope can be operated remotely using the appropriate software.*

NASA Addresses
http://spacelink.nasa.gov/index.html
http://www.hq.nasa.gov/office/pao/History/history.html
http://spacelink.nasa.gov/CORE/

NASA (the National Aeronautics and Space Administration) provides loads of "good stuff" on current and past space missions, aeronautics, and space-related topics. The first address has a search engine and gives you entry to NASA's extensive information files, the second address specializes in NASA's history, and the third address gives you a source for buying NASA materials.

Nine Planets

http://seds.lpl.arizona.edu/nineplanets/nineplanets/nine planets.html

The Nine Planets is an overview of the history, mythology, and current scientific knowledge of each of the planets and moons in our solar system. Each page has text and images, some have sounds and movies, most provide references to additional related information.

Saturn and the Saturnian System—The Cassini Voyage to Saturn

http://www.jpl.nasa.gov/cassini/

This page houses information and pictures concerning the Cassini voyage to Saturn and the Saturnian system. This is just one small example of the type of links you can find at the NASA addresses earlier.

Space Educators Handbook

http://tommy.jsc.nasa.gov/~woodfill/SPACEED/ SEHHTML/

Space comics (very cool), fun facts, and astronomy information all make up a part of this site. It is another example of the "good stuff" you can get to from the NASA addresses given earlier.

Students for the Exploration and Development of Space (SEDS)

http://www.seds.org/
The astronomy links at this site have particularly useful information.

Views of the Solar System

http://bang.lanl.gov/solarsys/
Although some of the information is written quite technically, this page does an excellent job of presenting the planets. The pictures and layout are great. There are lots of images and even some short films.

Botany and Agriculture

Future Farmers of America

http://www.agriculture.com/contents/FFA/index.html
You can find agricultural career information, state chapter home pages, and agricultural news at this site.

Environmental

Envirolink Network

http://www.envirolink.org/aboutel/
Envirolink bills itself as the best source of environmental information on the Internet. That just might be true! You will find instructions on how to express your views to your congressman and plenty of reference material to back up your points.

Environmental Education Network

http://envirolink.org/enviroed/

This site is valuable for its own environmental education resources as well as its links to other educational sites.

Environmental News Network
http://www.enn.com:80/

With news from a number of major wire services, this site provides not only the hottest stories and product information, but a good library as well. There is, however, a $25 yearly membership fee.

Greenpeace World Wide Web
http://www.greenpeace.org/index.shtml

Greenpeace, best known for its aggressive environmentalism, has built a wide-ranging site for friends of the Earth. You get news bulletins, instructions on taking action, and a comprehensive library. A WebChat feature keeps you in touch with Greenpeace supporters around the world.

Science and the Environment
http://www.cais.com/publish/

This bimonthly on-line publication has current news summaries on the environment that include colorful photos and graphics.

Geology (Including Earth Science and Weather)

Live from Antarctica
http://quest.arc.nasa.gov/antarctica/

Here is news of basic research on surely one of the most unique (and difficult-to-live) places on Earth. Find out what is happening!

Mesoscale and Microscale Meteorology for Information on Weather

http://www.mmm.ucar.edu/mmm/home.html

This site features some of the most up-to-date research for understanding atmospheric phenomena on spatial scales ranging from micrometers to megameters and timescales from seconds to a few days. You also can find some fascinating links; for example, one link <http://www.sover.net/~kenandeb/fire/hotshot.html> has a whole photo journal of U.S. Forest Service firefighters battling blazes in America's wildernesses. MMM (Mesoscale and Microscale Meteorology) is one of five science divisions in the National Center for Atmospheric Research, which is sponsored by the National Science Foundation and managed by the University Corporation for Atmospheric Research. The MMM Division is located at NCAR's Foothills Laboratory, 3450 Mitchell Lane, Boulder, Colorado.

Mount St. Helens

http://volcano.und.nodak.edu/vwdocs/msh/msh.html

Found here are pictures of and information on Mount St. Helens before, during, and after the volcano erupted.

Museum of Paleontology

http://www.ucmp.berkeley.edu/

Want to find out more about fossils? Here is one place you might want to start.

New York Earth Science Teacher

http://ourworld.compuserve.com/homepages/CVisco/

This site has hints for projects as well as links to earth science sites, specifically sites on rocks, mineral, and fossils.

Volcano World

http://volcano.und.nodak.edu/

Interested in dramatic happenings? Check out this site!

Weather Channel Online

http://www.weather.com/homepage.htm/

What's the weather? Here you can find current conditions and forecasts and view national weather maps, tropical storm and hurricane graphics, and surface weather maps.

Health and Medicine

History of Medicine Division

http://www.nlm.nih.gov/hmd/hmd.html

Need a historical perspective on a health topic? Try here!

The Heart Preview Gallery

http://sln2.fi.edu/biosci/preview/heartpreview.html

Doing a project on the circulatory system or the heart? Here is a good place to start for background information.

Physics and Chemistry

About Rainbows

http://www.unidata.ucar.edu/staff/blynds/rnbw.html

Did you ever hear of the moon making a rainbow? It is rare but it does happen! This tidbit of information as well as in-depth explanations of rainbows can be found at this site.

Microworlds—Exploring the Structure of Materials
http://www.lbl.gov/MicroWorlds/
Microworlds is an interactive tour of current research in the materials sciences at Lawrence Berkeley Laboratory's Advanced Light Source (in Berkeley, California). The Advanced Light Source is a pretty amazing machine. This site shows you what the ALS looks like inside, what it is like to work at the laboratory, and how infinitesimal quantities of trace elements can change a material for better or worse. It has an environmental connection by explaining how materials science is helping us understand environmental problems.

Optics for Kids
http://www.opticalres.com/kidoptx.html
Here you can learn about light, lenses, and more. The site contains a list of optics-related topics, and you can choose whichever one you need.

General Science

Nye Laboratories Online
http://nyelabs.kcts.org/
If you already know who Bill Nye the Science Guy is, I bet you will try this address. If you do not know who this character is, be sure to find out by accessing this site—you are in for a really fun discovery!

Exploratorium Home Page

http://www.exploratorium.edu

This site is full of electronic software that can be downloaded. It is an entertaining way to look at astronomy as well as physics and chemistry. This site also has a listing of 22 categories with over 250 annotated sites. The famous museum The Exploratorium is in San Francisco, California.

Great Lakes Science Center

http://www.greatscience.com/

Try your luck on an animated quiz about the Great Lakes, catch a piece of an OMNIMAX Theater film, find some experiments to do at home, and check out a few of the center's intriguing exhibits. The Great Lakes Science Center is in Cleveland, Ohio.

Nobel Prize Internet Archive

http://www.nobelprizes.com/

Annotated lists of all Nobel laureates in physics, chemistry, literature, peace, economics, and physiology and medicine are listed here, starting with the most recent and going back year by year. In 1912, for example, Swedish scientist Nils Gustaf Dalen received the Nobel prize in physics for his invention of automatic regulators for use in conjunction with gas accumulators for illuminating lighthouses and buoys. That is a mouthful, but isn't it intriguing that an invention for lighthouses in 1912 won a Nobel prize?

University of California Berkeley Center for Science Education

http://cse.ssl.berkeley.edu/

Check out this site for its Spanish-language programs, ready-made activities, and a link to Science On-Line (SOL).

Zoology

Birmingham Zoo

http://www.birminghamzoo.com/

This site features pictures and facts about animals. You can even "Ask the Zookeeper"!

SeaWorld /Busch Gardens

http://www.seaworld.org/infobook.html

You can search for fun facts on this page! There is information on many different animals.

Index